Mind-Boggling

Astronomy

STEVEN R. WILLS

"Reaching the moon by three-man
vessels in one long bound from Earth is
like casting a thin thread across space.
The main effort, in the coming decades,
will be to strengthen this thread, to
make it a cord, a cable, and, finally, a
broad highway."

Isaac Asimov, "The Coming Decades in Space,"
The Beginning and the End, 1977

Thanks to the following sources of inspirational models for adapted illustrations in this text: Page 25, adapted courtesy of *Coming of Age in the Milky Way,* by Timothy Ferris, published by Wm. Morrow & Co. Page 114, adapted courtesy of *A Teacher's Companion to the Space Station,* published by The Louisiana Nature & Science Center. Page 133, courtesy of NASA Publications. *From the Big Bang to Planet X,* by Terence Dickinson, published by Camden House and illustrated by Margo Stahl: Page 128, adapted courtesy of Terence Dickinson. Page 140, adapted courtesy of the Space Telescope Science Institute, a division of NASA, illustrated by Dana Berry. Page 147, adapted courtesy of Margo Stahl.

Copy-edited by Phyllis Lindsay
Design and page layout by Ann Turley
Cover illustrations by Mark James
Interior illustrations by Mark James
Technical illustrations by Sandra J. Joncas
Printing and binding by D.B. Hess Company

Manufactured in the United States of America

ISBN 0-942389-11-5

INTRODUCTION

by Paul Zaloom, "Beakman" of Beakman's World

Ahhhhh...the cosmos! As our global village shrinks by the day, the last frontier is beyond the clouds. Look up to the stars and see our universe. And, man, it is weird out there. What can we learn about ourselves by looking into space? I'm not a big fan of astrology; that's not what I have in mind. Instead, all kinds of interesting questions arise about our human place in the Big Picture.

What do we know about the origins of our universe? Is the Earth really hurtling through the universe at a speed of approximately 2 million kilometers per hour? In the time it takes you to read this sentence, will the universe really expand by 500,000 kilometers? Is it true that some of the dust on the furniture in your bedroom might be from space? How do astronauts go to the bathroom in their weightless environment without making a horrible mess?

Here's where you'll find the answers. And this book is very appropriately titled "Mind-Boggling," because what's in here is indeed a noodle bender. A certain wild and weird wacko named Steven Wills decided that astronomy needed a new take, a new interpretation, to get *you* involved and interested. Hey, it worked on me. So, sit back, take a deep breath, strap on your seat belt, and relax as you are hurtled into the mysteries, the wonders, and the just plain fun of the cosmos. Badda bing, badda boom, badda Big Bang!

FOREWORD

Mind-Boggling Astronomy is not designed as an astronomy text, but rather as a supplement to the enjoyment of science.

Why astronomy? We have always looked to the frontier with excitement and wonder, and while there are many frontiers remaining for our species (the oceans, the worlds of medicine, and human nature itself come to mind), space is most tangibly recognized as the "wilderness" of the next millennium.

Why now? We seem poised at an astronomical junction. Our memories of the race to the moon are fading into history, and the last 20 years of space exploration seems littered with half-hearted projects. The 1990s, however, may offer a glimmer of re-awakening. Today, there are three reasons to assume that our will (and need) to explore space has been rejuvenated. First, the Hubble Space Telescope is fueling astronomical research with so much data that scientists are unable to keep analysis at the same pace as input. Second, our images from the Voyager missions were so spectacular and so vivid in our minds that we see space once again as a place of incredible wonder. Third, our awareness of the fragile nature of Earth's ecology has placed a new urgency upon the study of Earth as a global unit, a "living planet."

Mind-Boggling Astronomy is intended to be full of surprises. We can't forget that science (and especially astronomy) should be fun. A sense of humor and a touch of irreverence does not destroy a topic, it enhances it. The current crop of fast-paced, in-your-face science shows on television indicates a renewed recognition that science can be fun, and that learning can be filled with the same joy as accompanies any exploration.

The classroom teacher should take special note of several additions to the book. Each sec-tion lists appropriate vocabulary, and a full glossary can be found in the back of the book. Each section also includes a "Teacher's Companion," listing discussion or writing topics, additional activities, and puzzle solutions — with particular emphasis on multi-disciplinary topics and cooperative learning situations. An activities index accompanies a comprehensive general index at the back of the book. There is also a section that lists further reading titles, software and on-line services, and organizations that specialize in astronomy activities for school-age children. While these additions are designed for classroom use, the individual reader or home-school user will find them equally helpful.

Mind-Boggling Astronomy was a project that would have been entirely beyond my scope without the assistance of: *Cobblestone*'s assistant publisher and editor, Carolyn Yoder, who maintained her faith despite the crumbling of deadlines; my English classes, who pretended not to notice that I was not always giving them 100 percent of my attention; astronomer and consultant Jim O'Leary from the Maryland Science Museum, who was gentle when noting where my science was shaky; illustrators Mark James and Sandra J. Joncas, whose research and blend of accuracy and lunacy provided a perfect backdrop for this book; and my wife, Susan, who has served as cheerleader, psychologist, hard-nosed editor, creative consultant, researcher, proofreader, organizer, motivator, instigator, prodder, and friend.

ACTIVITIES LISTING

This is a complete list of activities located throughout the book. Additional activities are located under Teacher's Companions at the end of each section. (Mind-bogglers are scattered throughout the book. In other words, find those yourself!)

SECTION 1

Create a Myth	11
Zodiac for the 1990s	11
Redesign the Calendar	15
Alien Calendar	15
Ellipse This!	21
Kepler's Kunfusion	22
Newton's First Magic Trick	26
Newton's Action-Reaction	27

SECTION 2

Measuring Distant Objects	36
Using Your Hand as a Sundial	37
Mapping the Lunar Surface	41
Demonstrating the Moon and the Tides	42
Seeing Mercury	47
Seeing Venus	48
Understanding a Terrarium	50
Colonizing Mars	50
Demonstrating Centripetal Force	56
Charting Jupiter's Moons	57
Ultimate Solar System Activity Part 1	64
Ultimate Solar System Activity Part 2	66
Design a Life Form	68
Seeing Sunspots	68

SECTION 3

Demonstrating Earth's Magnetosphere	76
Demonstrating the Coriolis Effect	82
Demonstrating Plate Tectonics—Maybe	83
Model of Plate Tectonics—Maybe	84
Rock'n'Roll Double Puzzle	85
Designing a Solar Collector	88
Lifetime Activity	88
Building a Theodolite — Uh-huh	91
Colors of the Spectrum Using a Pen	97
Graphing Space Photographs — Sort Of	99
Capturing Infrared Redhanded	100
Building a Cheap Rocket	102
Make Your Own Message from Earth	105
Can You Extrapolate?	106
Redesign a Space Shuttle Menu	109
Replacing Gravitational Force	112
Constructing a Space Station — Sort Of	114
Crossword Puzzle of the Stars	115
Feedback Data From Owners of Omicron IV	116

SECTION 4

Birth of a Star: Balloon Trick No. 1	126
Locating Famous Stars	128
Demonstrating Escape Velocity (Get Out of Town!)	129
Making a Black Hole — Sort Of	130
Graphing the Local Group	134
Moving Galaxies: Balloon Trick No. 2	135
Demonstrating Doppler	136
Gravity Hangs in There	142
Dial the Evolution of the Universe	143
How Many E.T.s Are Out There?	148

CONTENTS

INTRODUCTION   3

FOREWORD   4

ACTIVITIES LISTING   5

SECTION 1

THE BEAR NEVER BATHES

Astronomy Before Einstein   8

The Terminator   9

The Myth-Makers and Storytellers   9

The Observers   12

Year-Round Records   14

The Big Five (How to Form a Basketball   16
    "Dream Team" from Early Astronomers)

Claudius Ptolemy   17

Nicholas Copernicus   18

Johannes Kepler   19

Galileo and Isaac Newton   23

Teacher's Companion   28

SECTION 2

IT'S A BEAUTIFUL DAY IN THE NEIGHBORHOOD

The Solar System   32

The Terminator II   33

I've Got the Sun in the Morning...   33

...And the Moon at Night   38

Rock On! (The Terrestrial Planets: 42
    Mercury, Venus, Mars, and a Little
    Bit of Earth)

Terraforming Interruption   48

The Planetary Tour   50

The Asteroid Belt   50

Jupiter   53

Saturn   57

Uranus and Neptune   60

Last Stop: Pluto and the Kuiper Belt   62

Teacher's Companion   71

SECTION 3

OPERATIONS MANUAL

Spaceship Prototype OMICRON IV   74

Warranty   75

Specifications   75

Interior   75

The Terminator   76

Surface   77

Atmosphere   80

Development   82

General Maintenance   85

Accumulation of Greenhouse Gases   85

Depletion of the Ozone Layer   86

Parts and Service   89

Telescopes and Observatories   89

Telescope Locations 94

Cameras and Imaging 96

Rockets and Robots 100

Our Interstellar Travelers 103

Service Representatives 104

Accessories 106

Futurism 106

Living in Space 106

The Space Shuttle Experience 107

The Space Station Stop-Over 111

Space Spin-Offs 111

Colonization Is for Keeps 113

One Last "Oh Wow!" 116

Teacher's Companion 117

SECTION 4

LET THE DRAMA BEGIN!

Cosmology And The Questions Of The Universe

Cosmology And The Questions Of The Universe 122

The Terminator 123

The Age of Newton Has Ended 123

Enter Albert Einstein 123

Stars Are Born and Stars Die 124

Black Holes: The Universe's Way of Saying "Gimmie" 128

Galaxies: Two's Company, 200 Billion Is a Crowd 131

The Big Bang: The Ultimate Birthday 137

The Cosmic Timeline 139

Dark Matter 142

Time: Can I Get There Yesterday? 144

Wormholes: Where Space Equals Time 146

So Where Do We Go From Here? 146

Teacher's Companion 149

ADDITIONAL READING 152

GUIDE TO INFORMATION AND RESOURCES 152

GLOSSARY 155

CHARTS AND TABLES 156

BIBLIOGRAPHY 158

INDEX 160

Section 1

THE BEAR NEVER BATHES

Astronomy Before Einstein

When *did* astronomy begin? If you were going to hop into your personal space shuttle and slam-dance with a time warp, would you find the start of astronomy in the year A.D. 1609, with Galileo's construction of his first telescope? Or did astronomy begin in A.D. 1232, when the Chinese freaked the Mongols with rocket arrows? How about 270 B.C., when Greek mathematician Aristarchus described the Earth and other planets as revolving around the sun? (Talk about being ahead of your time! That was more than 1,800 years before Copernicus tried *again* to convince folks that the Earth wasn't at the center of the solar system. Sheesh!)

Did astronomy begin in 763 B.C., when Babylonians recorded a solar eclipse? Or 2296 B.C., when the Chinese described sighting a comet? Or how about 6500 B.C., with the earliest known calendar — carved on bone by a primitive resident of Zaire? Or how about 7221 B.C., when Orag gazed over at his cave-wife and said, "Honey, can we invent some astronomy tonight?" (Only kidding.)

It's impossible to find the beginning of astronomy; but one thing is clear. Astronomy is the oldest science. No doubt about that.

THE TERMINATOR

You can check these out in the "Glossary" in the back of the book.

solar eclipse
hypothesis
astrophysics
constellation
horoscope
lunar eclipse
supernova
sunspot
Heel stone
solstice
medicine wheel
geocentric theory
scientific method
heliocentric theory

precession
magnitude
epicycle
dogma
gravity

Enough already! I'm terminated!

THE MYTH-MAKERS AND STORYTELLERS

It's not hard to see why astronomy is such an ancient science. After all, when primitive men and women first looked to the sky and wondered — couldn't they have been called astronomers?

One way to focus on the origin of astronomy is to think of what astronomy *is* rather than when it began. Astronomy, if you want to strip it right down to its underwear, is applying what we *know* to solve the mysteries of what we *don't know*, particularly as they relate to our universe. (Remember that, it'll probably be a quiz question.)

For example, even mythology can be a form of astronomy.

So how can myths be astronomy? Isn't astronomy a kind of science and myth a kind of religion?

True, but religion and science aren't always that different.

Need proof? No problem!

When ancient Egyptians saw the sun rise in the east, move across the sky, and set in the west, that was a mystery. Just what were they supposed to make of it? The sun moves, right? It looks and feels like a big round fire, right? It disappears at one horizon and reappears the next morning at the other horizon, right?

Okay, then, how could these Egyptians use what they *knew* to explain such a strange thing?

Their thinking might have gone like this: The sun moves smoothly, like a fire-boat sailing on blue water. This water is in the sky, however, so it must be a heavenly ocean. There must also be some place where the fire-boat sails at night, so that it can appear once again in the east. Such a boat of fire must be piloted by a being greater than human. Their conclusion: The sun god, Ra, sails from east to west each day, providing light to view his subjects below. In the evening, Ra sails to the underworld until the start of a new day.

Myth? Yes. Astronomy? Well ... yeah.

In cultures around the world, people looked to the sky and created stories to explain such things as phases of the moon, changing of the seasons, high and low tides, stars and comets, and much, much more. These stories became a part of tradition, of culture, and of religion.

HEY JULIUS, BETTER LATE THAN NEVER!

Julius Caesar is given a lot of credit for organizing the 365-day calendar and adding a leap year. That was in the year 46 B.C. However, the first 365-day calendar really was developed by the Egyptians back in 2773 B.C.

And if you were a farmer trying to predict when the rainy season would begin, the stories were also a part of your science.

Some of our best known myths come from the Greeks and Romans. For example, we know that the seasons change for astronomical reasons. The Greeks, however, sorted it all out like this:

Demeter was the goddess of growing things. She had an affair with Zeus (who, even though he was married to Hera, still managed to have an affair with pretty much anybody who wore a toga-skirt). Their daughter was called Persephone — the maiden of the spring.

Persephone must have been beautiful, because Zeus' brother Hades, lord of the underworld, fell head over Achilles-heels for her and kidnapped her. And even though Hades was in charge of the world's riches (gold comes from under the ground, right?), Persephone was not thrilled with living in this dark underworld, and her mother, Demeter, went absolutely ballistic. She declared that nothing would grow until Persephone returned.

Zeus, however, couldn't let this go on, and worked a deal. Persephone could stay with Hades for four months of the year, but would be allowed to return to the surface world for the other months. When Persephone was with Hades, Demeter would not allow anything to grow (winter). When Persephone returned, spring arrived. In the autumn, Demeter and Persephone began to weep for her coming return to the underworld, and all living things would begin to shrivel and die.

And that, according to the ancient Greeks, was why the seasons changed.

Native American myths follow the religious belief that all things possess a spirit and life-force. These myths don't usually involve human-like gods representing natural forces (such as Apollo pulling the sun in a chariot), but the forces themselves as living beings (the

Sun Father). Also, Native American myths are loaded with references to the natural wilderness phenomena that are so much a part of their culture.

In one story, the Sun Father goes searching for his sister, called Night Sun. (The Night Sun is the moon.) She had become lost in the path of the stars. As he searches, the Sun Father sees waterfalls of stars and the Great Bear, which, long ago, had been wounded by hunters on Earth. (The myth describes how the blood of the Great Bear each year paints the leaves with red.)

When Sun Father finds Night Sun, she is nearly starved. Sun Father builds up her strength, fattening her once more. But then, she sees herself restored to fullness, becomes alarmed, and runs off again. Every month, Sun Father finds and fattens Night Sun. Every month, Night Sun runs off and nearly wastes away from starvation.

The phases of the moon.

2,200 YEARS BEFORE THE PSYCHIC CHANNEL ARRIVED ON CABLE TV

By the year 400 B.C., the Babylonians had charted the 12 constellations of the zodiac (the zone of the sun's annual path). Horoscopes based on birth dates also were available (although probably not from vending machines).

Keep in mind, it wasn't so long ago that some scientists thought they observed "canals" on Mars and formed many hypotheses to explain the phenomenon — some involving entire civilizations. At the time this seemed like scientific theory. Later knowledge revealed these stories of Martian civilizations to be pure fiction — a kind of myth.

And keep this in mind, too: Today most scientists believe that objects cannot travel faster than the speed of light, based on what we *currently believe* to be the laws of astrophysics. Might we one day learn we were wrong? Could our belief turn out to be more like myth than reality? Probably not; but who knows, maybe those early Egyptians also said "probably not."

The point is, before we toss off some belief as unscientific, we have to understand — sometimes the evidence changes.

ACTIVITIES

CREATE A MYTH

(Don't panic, it isn't that hard.) Pick something you don't understand — something mysterious. For example, how does a television work? or is there an edge to the universe? or what happens inside a black hole? or where does all the lint come from that you dig out of your clothes dryer (or bellybutton)? Create characters (Fiberus, the god of lint?), add some conflict, and make a story.

ZODIAC FOR THE 1990s

Have your science teacher display the constellations that make up the signs of the zodiac. (He or she probably has charts of these someplace. Science teachers never throw anything away.) Forget the traditional names and figures, and use the same star groups to create a zodiac for the 1990s. (Do any look like the Shaq?)

Talk about ugly.... There was supposed to be a mega-neat section here about the constellations of the zodiac and some of the brightest stars appearing in those constellations — real prize-winning stuff!

Wouldn't you know it, no sooner does the copy get entered into our publisher's computer, than the computer operator (who was working through lunch hour because he fell asleep in the morning because of *Monday Night Football*, which is another story) decides to take a drippy bite of his pizza (which he ordered out so he could work through lunch hour...you get the picture). Well, the cheese is hot and suddenly he lets out this "GRREEK!" and a large pepperoni slips off his chin and lodges itself between two of the function keys.

Next thing you know, the lights flicker, there's a strange smell in the office, and — bottom line — the section on the zodiac is history.

Help!

In this section, we were going to discuss five constellations of the zodiac: Virgo, Gemini, Taurus, Scorpius, and Leo. We know that at 9 p.m. in the northern hemisphere two of these constellations appear in winter, two in spring, and one in summer. We also know that each of these constellations contains a widely known star: Pollux, Aldebaran, Spica, Regulus, and Antares.

Can you match the constellation to its star and its season? Here are some clues we found in notes near the toasted computer console:

1. Virgo and Regulus appear in the same season, but not at the same time. The same can be said of Gemini and Aldebaran.

2. Gemini and Taurus see the solstice, while Spica and Leo see the equinox.

3. Aldebaran arrives in the same season as Pollux, but after Leo.

By the way, don't bother looking at some newspaper astrology page to find a match between signs of the zodiac and months. We already tried that, and it doesn't work. Those matches were made long ago, and the precession of the equinoxes (the wobbling of the Earth's axis of rotation) has changed things.

Good luck, and if our former computer operator were still working here, he'd say, "Thank you."

THE **O**BSERVERS

Myth-making was not the only astronomical activity going on in the ancient world. There also was a good deal of scientific recording and observation. The Babylonians, for example, compiled probably the earliest record of astronomical events, dating back to 1800 B.C. They charted the stars, recorded a solar eclipse, and mapped out the zodiac (but you already knew that). The Babylonians observed and recorded for more than 1,000 years, until around 700 B.C., when the Assyrians torched their city. (The

Babylonians became powerful again, however, and again began a scientific investigation of the heavens…until 539 B.C., when the Persians torched their city. Bummer!)

The Phoenicians noticed that, throughout the night and throughout the year, one star seemed to remain motionless in the night sky. This star, Polaris, was circled by the widely known Great Bear constellation Ursa Major (The Big Dipper). At the latitude of the Mediterranean, the Great Bear never sank below the horizon. The Phoenicians even made up a phrase to remember this important fact: "The bear never bathes." (Of course, they said it in Phoenician.) Armed with this knowledge, the Phoenicians were able to attempt something no earlier Mediterranean culture had attempted.

They sailed beyond the sight of the shoreline.

They sailed by the stars.

The Chinese also were terrific observers. From them, we have the first recorded comet (2296 B.C.), the first recorded lunar eclipse (1361 B.C.), and the first recorded supernova (352 B.C.).

THAT'S BECAUSE NOBODY TORCHED THEIR CITY

In 165 B.C., the Chinese recorded the first sunspots. They kept a continuous record of sunspots from 28 B.C. all the way to A.D. 1638.

*R*ecords *I*n *S*tone

Some sky-watchers recorded events on papyrus, some on animal hides, and some on stone.

And some recorded events *with* stone.

Stonehenge (like the Sphinx, the Mayan pyramids, and your math teacher) strikes us with its age, its size, and its mystery. The massive

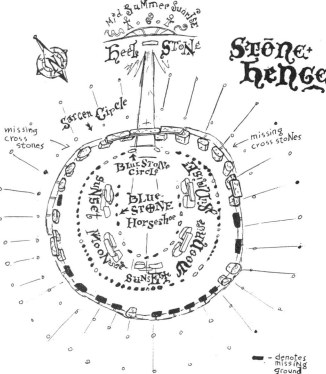

stones still stand in the morning fog of the English countryside, a ghostly reminder of distant times and strange civilizations.

Stonehenge, from what we have been able to decipher, is part religious temple and part astronomical observatory, and has stood as such for nearly 5,000 years.

Stonehenge was constructed in three phases. Phase I was built nearly 5,000 years ago and consisted of mounds of rock. One stone, called the Heel stone, was set at this time and is the most important, astronomically. If you stand in the center of the circle and look to the Heel stone, you look directly to the point on the horizon where the sun rises on the summer solstice.

Why was this date so important? Well, on one hand it was important to know when daylight would stop lengthening and begin to shorten. But even more than that, for the early skywatchers it was important just to know that such a pattern existed. It would be a pretty frightening world if you believed the sun could rise and set whenever it felt like it!

Other "station stones" (stones that serve as markers) set during Phase I mark where the moon rises and sets on key days in its cycle. Clearly, the builders of Stonehenge were careful skywatchers.

Phase II construction is not clearly dated, and seemed more concerned with ritual and may mark the use of Stonehenge as more of a temple than an observatory. Larger stones were

SPEAKING OF BIG STONES, AND HOW TO IMPRESS PEOPLE

The Great Pyramid at Giza is aligned to Polaris, the North Star. (Don't tell anyone you know this. Then, next time you're in Egypt, you can look over at the Great Pyramid in a casual sort of way and say, "Gee, isn't that aligned to the North Star?")

brought to the site and placed to permanently mark the circles from Phase I.

Phase III, accomplished sometime around 1600 B.C., involved setting the huge "sarcen stones." These are the large stones seen in photographs of Stonehenge. This must have been one heck of a project, since they weigh up to 50 tons each. It is estimated that about 1,100 men were needed to haul them, on wooden rollers, from a quarry 25 kilometers away. The trip would have taken seven weeks for each stone. And if that wasn't enough to impress primitive chiropractors, the "lintel stones" then had to be set across the tops of the standing sarcen stones.

UN-BOGGLE THIS!

Nobody knows for sure how these lintel stones were raised and set in place. They were lifted six to eight feet and placed across the sarcen stones without the help of any technology (remember, this was 1600 B.C.) The lintels also weighed up to 50 tons each.

Now let's say you have been given the job of getting these stones into place. You have a good supply of laborers, but not much else. You do, however, have a mind-boggler's imagination. Develop a plan that would get the job done. If you want to show off, develop two plans.

So who built Stonehenge, and how, and why? Hey, it's a mystery! The point is, some culture back in the early Bronze Age knew enough about the sun and stars to create an observatory. And they knew enough about how important astronomical events were in their lives to build that observatory so it would last.[*]

* Renfrew, Colin. *Before Civilization*. New York, NY: Alfred Knopf, 1973, pages 216–242.

YEAR-ROUND RECORDS

Most early cultures kept a calendar of celestial events. The best time to plant crops or expect harvests was vital information, as were the expected rainy and dry seasons. For nomadic cultures, the sky-watchers could predict the time to move to other areas in order to catch up with migrating herds of game.

Unless you had a calendar, you weren't playing in the big leagues.

The Plains Indians of North America prepared stone "medicine whe   as their calendars. Stones marked the po..ts where bright stars rose from the horizon on key dates, indicating when to migrate to seasonal grazing lands and when to plant crops.

The early Greek poet Hesiod offered this advice to Mediterranean farmers:

"When great Orion rises, set your slaves
*   To winnowing Demeter's holy grain*
Upon the windy, well-worn threshing floor.
Then give your slaves a rest; unyoke your team.
But when Orion and the Dog Star move
Into mid-sky, and Arcturus sees
The rosy-fingered Dawn, then Perseus, pluck
The clustered grapes, and bring the harvest home."[*]

The whole poem is the length of a book, but you get the idea. (The Greeks could have used a good wall calendar.)

However, nobody, *but no-body*, did calendars like the Aztecs.

The Aztec calendar was divided into 18 months, each month lasting 20 days. (Yes, that totals only 360 days, but hang on already!) Each month was named for something significant during that month. For example, the first month was usually dry and was called Atlcoulaco, the "want of water." The third month was Tozoztontli, or "fasting." Other

* Hesiod. *Works and Days*. Dorothea Wender, translator. New York, NY: Penguin Books, 1977, page 78.

DO NOT TRY THIS AT HOME

Several months in the Aztec year were celebrated with human sacrifice. As a matter of fact, it was considered good luck on such celebrations to dance while wearing the skin of your sacrificial victim.

What do you think of your lucky rabbit's foot now?

month-names included "eating the corn," "birth of flowers," "falling fruits," "falling waters" — finally the 18th month, Izcalli, the time of mass sacrifice (a good month to be out of town).

Okay then, what about those extra five days? After all, they can't be ignored or, over a period of years, the month of falling fruits will arrive before the fruits fall — and you wouldn't want to be the Keeper of the Calendar during Izcalli of that year.

ZERO HOUR

The Christian calendar (also called the Julian calendar) begins with the birth of Jesus Christ. Other calendars have different starting points.

Sometimes the simplest solution is the best. The Aztecs

The Roman calendar begins with the founding of Rome in 753 B.C.

The Islamic calendar begins in A.D. 622 when the prophet Mohammed left the city of Mecca.

The Jewish calendar begins with the supposed creation of the world in 3761 B.C.

The Greek calendar begins with the first Olympic games in 776 B.C.

Sooo... 👉 *continued*

☞ The Christian year A.D. 2000 would also be
the Roman year 2753
the Islamic year 1378
the Jewish year 5761
the Greek year 2776

regarded those extra five days as bad luck — after all, they messed up the "elegance" of their 18 equal months. So they did nothing.

Nothing!

The five "empty days" were called Nemontemi, and they carried such bad vibes that no one was allowed to dance or sing or start a fire or anything. All you could do was hunker down and wait — hoping the bad luck didn't have your name on it.

And what about leap year? That's not entirely clear, but occasionally the five Nemontemi probably just became six Nemontemi. Extra bummer!

ACTIVITIES

REDESIGN THE CALENDAR

Throw out all the rules and design your own calendar. Start your year whenever you think best and divide it up in your own way. What's your plan? (You do need a plan.) How is your calendar better than the cheesy ones we've been using for all these years?

ALIEN CALENDAR

Design a calendar for another planet, or even for our moon. Research and find out periods of rotation and revolution (or peek ahead in this book), and develop a yearly chart that makes some sense. Hey, someday there may be a colony on the planet you choose, and they may end up using your calendar. We're talking fame here, not to mention product endorsements.

WHERE NO BOGGLER HAS GONE BEFORE...

Anybody out there a Trekker?

If you are, then you're familiar with the "stardates" used in both *Star Trek* and *Star Trek: The Next Generation* television series. Writers and producers of the shows have been careful to make those stardates believable so that they fit a certain chronology. A careful viewer also can figure out when certain things happen in "Earth time."

For example, Captain Kirk was born in the year 2233 , and Captain Picard was born in the year 2305 (Earth time).

Kirk's first mission on the *Enterprise* ("The Corbomite Maneuver") was in the year 2266 Earth time, but was also stardate 1512.2.

Picard took over the rebuilt *Enterprise* ("Encounter at Farpoint") during Earth year 2364, which was stardate 41153.7.

Not only that, the movie *Star Trek: the Motion Picture*, takes place in the year 2271 and at stardate 7412.6.

So, then, the following dates match up:

Earth Year	Stardate
2266	1512.2
2271	7412.6
2364	41153.7

For years, fans have tried to figure out a formula to convert Earth years to stardates.

Give it a try!

THE BIG FIVE

(How to Form a Basketball "Dream Team" from Early Astronomers)

WARNING: The publishers of this book want to point out that the facts mentioned in the interviews that follow are true — even the ones about Tycho Brahe's nose. However, they do not believe there has ever been a basketball team formed from dead astronomers. On the other hand, they probably weren't watching cable last Wednesday at 3 in the morning.

Network Special

We interrupt this terrific astronomy book for the following special program:

"Well, Bud, the coliseum is beginning to fill up now, and I can't imagine a greater show than the one we're about to see. This Dream Team has been gathered from history and represents the finest astronomical talent ever brought under one roof."

"That's right, Al, and no one epitomizes that talent more than Claudius Ptolemy himself. It's true that Ptolemy has taken a lot of heat lately for some bad judgment calls on his part; but I have to say, I think the whole controversy has been blown way out of proportion."

"I couldn't agree more, Bud. That's one thing our up-close-and-personal reporter, Miles Evergrin, spoke to Ptolemy about recently. How about if we roll that tape?"

*C*laudius *P*tolemy
(A.D. 100–170)

Miles Evergrin: Astronomy fans, we're here today in Alexandria, Egypt, to visit with that great observer and astronomer, Claudius Ptolemy. Now, Claudius, let me begin with the question I know your fans want me to ask. When you published your geocentric theory in A.D. 150, well, how can I put this diplomatically…Earth at the center of the universe? What was going through your head?

Claudius Ptolemy: Quite simple, Miles. I thought I was correct. But I think some background is in order.

Back in the first century, Aristotle was the number one guy. My disagreements really began with him. Aristotle believed the Earth was the center of all the universe, and the sun and stars moved in spheres around the Earth.

M.E.: But Claudius, isn't that what *you* said, also?

C.P.: My disagreement was not over the geocentric theory; it was even more basic than that. Aristotle was a philosopher, and he didn't give a hoot for careful observation. My observations of the stars and planets showed that Aristotle's ideas were much too simple. Aristotle's universe didn't match scientific observation — the "evidence," you would say. My position was that the *observations must be preserved.* Evidence is more important than theory — and when the theory doesn't match the evidence, then the theory must be changed.

M.E.: I see. I suppose we could say that you were applying the scientific method a thousand years before it was generally recognized. But Claudius, didn't that Greek mathematician Aristarchus develop a heliocentric theory — with the sun at the center — way back in 270 B.C.?

C.P.: What can I say, Miles? I never paid Aristarchus much notice. Then my work, *Almagest,* was published in A.D. 150, and that was that. Sometimes I wonder how I made the Dream Team at all.

M.E.: Don't be modest, Claudius. Our fans certainly remember the accuracy of your predictions of the motions of the moon, the sun, and the stars. Your charts of the stars brought new definition to the idea of astronomical observation. It's no mistake that "almagest" is Arabic for "the greatest."

C.P.: But I was off. For heaven's sake, Miles,

> ### "Animals, which move, have limbs and muscles; the Earth has no limbs and muscles, hence it does not move."
>
> *Scipio Chiaramonti, Professor of Philosophy and Mathematics at the University of Pisa, 1633*

my calculations predicted the entire *universe* to be about 50 million miles in radius. Heck, the Earth is farther from the sun than that.

M.E.: Okay, but for your time, you were the best. The best — if not for your predictions of the motions of stars and planets, or for your catalog of more than 1,000 stars, then how about for your description of the wobbling of the Earth's axis — a phenomenon you describe as the precession of the equinoxes.

C.P.: Well, I "borrowed" that one from the early astronomer Hipparchus. But it *was* pretty cool.

M.E.: Darn right! And what about your descriptions of the brightness of stars? You designed the scale of "magnitudes" of brightness that we still use today.

C.P.: I, um…borrowed that from Hipparchus, too. This is embarrassing.

M.E.: Face it, Claudius, yours was not an easy time. By the end of your lifetime, civilization in Europe had begun to fall apart. The Roman Empire was collapsing. The gold and silver mines of Spain were exhausted, which ruined the world economy and froze trade. Not only that, the bubonic plague — the Black Death — took 2,000 victims each day. The barbarians were moving in, and the Dark Ages were about to begin.

C.P.: Yeah, things were looking pretty grim.

M.E.: Grim? The Dark Ages signaled the end of intellectual progress and nearly the end of humankind itself. It would be 1,300 years before the Renaissance brought science back to Europe. Claudius Ptolemy, you were the last great astronomer of the ancient world.

C.P.: Not bad, I guess.

"Well, that was some interview, eh, Bud?"
"Absolutely, Al. But we have someone in the studio right now who also has a pretty darn

good story of his own. It's none other than Niklas Koppernigk. Hey, glad to have you aboard, Nikki!"

*N*icholas *C*opernicus
(A.D. 1473–1543)

Nicholas Copernicus: "Please, call me Copernicus, Nicholas Copernicus."

Al: "Why the name change, big fella? The original spelling is, well, exotic — all those Ks and everything."

N.C.: "It may seem exotic to you, but I was born in central Poland, which wasn't exactly the center of the universe."

A: "Neither was the Earth, eh, big guy?"

N.C.: "Uh, yes. Anyway, I received the best education available at that time, so I did what most scholars did — I changed my name to its Latin equivalent: Nicholas Copernicus."

A: "I guess 'Elvis' was already taken, right?"
N.C.: "Huh?"

"Hey Al? Bud here. I'm not getting anything on my earphones. Are you there?"
"Why don't you check it out, Bud? Anyway, Nick, how about that heliocentric theory?"

N.C.: "It seemed to me, Mr. Al, that we needed to place the sun at the center of the universe. All this is suggested by the systematic procession of events and the harmony of the whole universe, if only we face the facts, as they say, 'with both eyes open.'"

A: "I, uh, understand, but, um, could you explain a bit more for our viewers at home?"

N.C.: "Sure. I surveyed the observations of Ptolemy and others — I never had the patience to be much of a stargazer myself — and realized that geocentricism could not explain the movements of the heavens. This was a problem, and the idea at length came to me how it could be solved — if some suggestions were granted me.

"First, there can be no one center of all the celestial circles or spheres.

"Second, the center of the Earth is not the center of the universe, but only of the lunar sphere.

"Third, all the spheres revolve about the sun as their midpoint, and therefore the sun is the center of the universe."

A: "So then, Nick, how do you feel playing on the same team as the father of geocentricism, Claudius Ptolemy?"

N.C.: "I have the finest respect for Ptolemy, Al. His observations were impressive, and I depended on them to form my own theories. And let's not forget that heliocentricism was not just my theory; Aristarchus voiced it 1,700 years earlier. And besides, I made a mistake or two myself."

A: "Mistakes? The great Copernicus? Come on, you're not fooling with our audience, are you Nick? After all, you described the moon's orbit and distance so accurately that we still use your formulas today. You also accurately described the Earth as a planet that rotates on its axis. You even recognized that Ptolemy's 'precession of the equinoxes' was the result of the Earth wobbling on its axis."

N.C.: "Yes, yes, very true, Al. But the picture wasn't yet complete. I tried to stick with the idea of circular orbits and epicycles, and all that. Circular orbits just wouldn't work. And comets! Let me tell you what kind of a mess *they* made of my theories. No, I knew there were problems. That's why I didn't publish my work, *De Revolutionibus*, until I was near death. I wanted the planets to move in perfect circles. The heavens, however, would not agree."

A: "Terrific story, Nick. It's great to have you on the team today. Any other questions, Bud? Bud?"

"Al? This is Bud. My headphones aren't working, and I can't hear anything."

"Okay, then, let's have a look at some footage from a recent profile of our next Dream Team player: Johannes Kepler."

*J*ohannes *K*epler
(A.D. 1571–1630)

tick, tick, tick, tick, tick......

Barely Safer: "Good evening. This is Barely Safer, and on this edition of *6 Minutes* we'll be talking with the brilliant mathematical astronomer who, despite having discovered the laws of planetary motion, remained, in the words of fellow astronomer Tycho Brahe, a 'squinty-eyed little loser.' Ladies and gentlemen, Mr. Johannes Kepler."

Johannes Kepler: "Brahe never said that about me. It wasn't easy working with that man, you know. Sure, he took volumes of notes on the movements of the stars and planets, but he ate like a whale, drank like a fish, and carried the belly of a walrus. The man strutted and yelled, and threw table scraps to his servant — a dwarf named Jepp. The man even had a metal nose after having his chopped off in a duel. And he thought I was a loser?"

B.S.: "Mr. Kepler, I was only…"

J.K.: "And do you know how Tycho Brahe finally kicked the bucket? One night in October at a royal dinner party, he drank so much beer his bladder burst. Poor hippo couldn't even tear himself away from the dinner table long enough to go to the bathroom. Brahe, humph — at least people know how to pronounce my name."

B.S.: "To be fair, Mr. Kepler, you are not known for your endearing personality, either."

J.K.: "Okay, okay, that may be true, Mr. Safer. I was always the sickly boy with the dog-like face, picked on and shunned. But I *did* make it, didn't I?"

B.S.: "Yes, you did. And now you've been selected for the Dream Team — to play alongside the likes of the great Galileo, Coperni—"

J.K.: "Galileo! Did you know that the 'great Galileo' laughed at me when I told him that the moon causes the tides? He laughed! 'Johannes,' he said, 'Poor Johannes. The tides are caused by the sloshing of the sea as the Earth spins on its axis.' The great Galileo…I never heard anything so silly in my life…the seas sloshing around."

B.S.: "Yes, I see. But getting back to your story, Mr. Kepler, could you take us back to 1600, and your arrival on the astronomical scene, as it were?"

J.K.: "Of course. I arrived at Benatek Castle

near Prague, Poland, in February 1600, where Tycho had moved his observatory. His previous benefactor, King Frederick, had recently drunk himself to death. Of course, Tycho found a new sponsor right away. Everything always came easy for him."

B.S.: "Yes, Mr., Kepler, and then?"

J.K.: "Well, Tycho had recorded mountains of observations — he was quite good at that, actually. I was trying to make some money selling horoscopes, but no one ever paid me.

"Anyway, the main problem at that time was figuring out how the planets moved. All the accepted theories involved circular orbits, and we spent most of our time trying to find little gimmicks to make the circles match our observations. It wasn't working.

"After Tycho died in 1601, it came to me. They weren't circles at all! The orbits of the planets are perfect ellipses.

"And better yet, once I accepted that orbits were ellipses instead of circles, my background in geometry brought other theories to the fore. These became my other laws."

B.S.: "Perhaps we should review for our audience, Mr. Kepler."

J.K.: "Of course, of course. Here it is, then:"

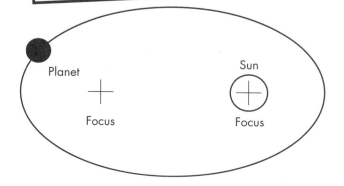

Kepler's First Law:
The orbit of a planet is an ellipse, with the sun at one of its two foci.

Kepler's Second Law:

In its orbit about the sun, a planet will sweep out equal areas in equal times. Therefore, when a planet is closer to the sun in its orbit, it will move faster than when it is farther away.

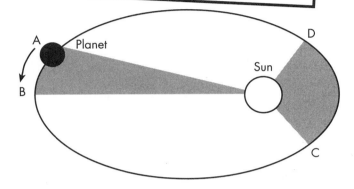

Kepler's Third Law:

The cube of the distance of a planet from the sun is proportional to the square of its period of orbit.

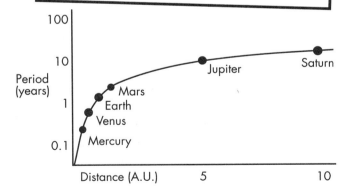

B.S.: "Truly amazing, Mr. Kepler."

J.K.: "Fat lot of good it did me, Mr. Safer! You would think that after a breakthrough like that, I'd have it made. But I couldn't get a break. My six-year-old son Friedrich died from smallpox. My wife's health faded until typhus took her. My mother was nearly executed as a witch and died shortly after her release. And to top it all off, I was released from my position as astrologer to Duke Albrecht von Wallenstein when I asked him to pay what he owed me. I was flat broke."

B.S.: "And yet, Mr. Kepler, you are now recognized as one of history's most notable astronomers."

J.K.: "Yeah! Just don't mention Tycho Brahe anymore."

tick, tick, tick, tick, tick, tick……

ACTIVITIES

ELLIPSE THIS!

For some reason, the concept of an ellipse bothers people. Ellipses, however, are easy to draw and examine.

YOU WILL NEED:

white paper
two push pins
a pencil
some string
a surface you can put the push pins into, such as a bulletin board or your younger brother or sister

(NOTE: The publishers of this book, their relatives, and certainly the author refuse to accept any legal responsibility for anyone so dim as to actually poke pins into a brother or sister and then call some fancy-shmancy lawyer looking for big bucks. Some people have no sense of humor.)

Tie a loop of string. Then push the pins through a piece of white paper into the bulletin board, so that there is plenty of slack as you drape the string around the pins. Place the pencil inside the string and pull it to the side until the string is taut.

Draw an ellipse, circling the two pins with the pencil while keeping the string taut.

The two pins are the two foci. In a planetary orbit, the sun will be one of those foci.

Use other pieces of paper to try changing some things. Change the length of the string. Change the distances between the pins. Observe the different orbits that result from these changes. They are *all* ellipses.

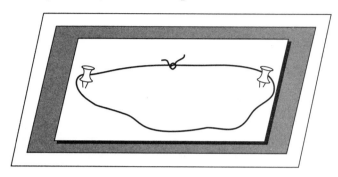

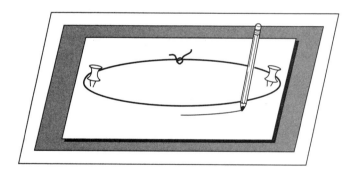

Kepler's Kunfusion

WARNING! What follows is not for the mathematically squeamish!

For this, you will need either a pocket calculator or a better head for numbers than most people.

Kepler's third law states that the cube of the distance of a planet from the sun is proportional to the square of its period of orbit. That sounds pretty confusing, but here's what it boils down to: The distance of a planet from the sun

determines how long that planet takes to revolve around the sun (its planetary year). The planetary year is often called its PERIOD, so we can set up a formula where P = Period and D = Distance (from the sun):

$$P1^2/P2^2 = D1^3/D2^3$$

That isn't much better. To simplify the formula, however, we can use a standard distance called an ASTRONOMICAL UNIT (a.u.). An astronomical unit is equal to the mean distance of Earth from the sun, so the distance of Earth from the sun is 1 a.u. The distance of Mars, for example, is 1.524 a.u. We'll also set a standard period based on Earth's year, so Earth's P = 1.

Now, with Earth as a standard, a little mathematical tinkering reveals a cleaner formula:

$$P^2 = D^3$$

Since $P^2 = D^3$, you could determine (even without a calculator) that, given that Earth's D = 1, then Earth's P also = 1 (Earth's period of revolution is one year).

Okay, then, if Mars' a.u. is 1.524, what is its period of revolution? (Take a deep breath and pull out the calculator.)

Here goes. If Mars' D = 1.524, then D^3 = 3.540. That means $P^2 = 3.540$, then P = 1.881. In other words, Mars takes 1.881 Earth years to orbit the sun. (Whew!)

If you're good (or if the people next to you aren't covering their answers very well) you can fill in the missing pieces for this table.

Planet	Period of Revolution (in years)	Mean Distance (in a.u.)
Mercury	?	0.387
Venus	?	0.723
Earth	1.000	1.000
Mars	1.881	1.524
Jupiter	11.86	?
Saturn	29.46	?
Uranus	?	19.18
Neptune	164.8	?
Pluto	?	39.52

(NOTE: Missing pieces: Mercury, 0.241; Venus, 0.615; Jupiter, 5.203; Saturn, 9.54; Uranus, 84.0; Neptune, 30.06; Pluto, 248.4.)

Mind-Boggler Alert!!

Your science teacher is too modest to tell you this, but he or she has just discovered a new constellation. This particular constellation is formed by five rows of stars. There are four stars in each row. The strange thing is, there are only 10 stars total in the constellation!

Can you figure out the shape of the new constellation?

...And Now, Back To The Show

"Well, Al, you might say we've saved the best for last. The final two members of our Dream Team. The Czars of the Stars."

"Yeah, Bud. The Twin Platoons of the Moons."

"The Double Tricks of Physics."

"Anyway, both of these great players are in the locker room, so why don't we just listen in as they get their 'game face' on for the Dream Team's appearance."

"—The Overjoyed of the Asteroid."

"Enough, Bud. We're going to listen in now to the pre-game conversation between Galileo Galilei and Isaac Newton."

* McAleer, Neil. *The Cosmic Mind-Boggling Book.* New York, N.Y. Warner Communications Co., 1982, page 116.

Galileo
(A.D. 1564–1642)

And

Isaac Newton
(A.D. 1642–1727)

Galileo: I don't know about you, Isaac, but these pre-game minutes are difficult. I find that the silence is filled with thoughts of my youth.

Newton: They are patterns, Galileo, my friend.

G: It seems the pattern of my life can be seen in those early years. Even in school I refused to accept dogma — the statements of those professors who expected me to believe them without proof. I challenged and questioned. After all, is our intellect to be enslaved by that of someone elses? And yet, perhaps I went too far. At the University of Pisa, they called me "the wrangler" for my argumentative sarcasm.

N: But that is also what brought you greatness.

G: And trouble. I even lost my position as professor of mathematics at Pisa because of my sarcastic lectures, questioning Aristotle's ideas. What about you, Isaac?

N: My childhood? Nothing much to say about that. I was an only child — unhealthy and not expected to live. My mother was widowed and remarried, and I was shuffled off to be raised by my grandmother. I was quiet — and boiling with anger.

G: Was there nothing bright in those years?

N: Oh, yes. I enjoyed tinkering and building. I built clocks and tried to make perpetual motion machines. I even studied plans for a universal language. There were bright moments, but they were mine alone. But Galileo, speaking of "moments," did you ever have an "Oh, wow!" moment?

G: An "Oh, wow!" moment? I think I know what you mean. Once, when I was just 18, I saw a lamp swinging on a cord from the roof of the Pisa Cathedral. I realized as I followed the cord to the roof, that, although the range of the swing decreased, the time for one swing remained the same. I suddenly knew I had found an accurate means of charting the passage of time. I tested this pendulum against a human pulse, and found it to be an accurate

> ### "All knowledge of reality starts from experience and ends in it. Because Galileo saw this, and particularly because he drummed it into the scientific world, he is the father of modern physics — indeed the father of modern science altogether."
>
> *Albert Einstein*

> ### "I think Isaac Newton is doing most of the driving now."
>
> *Astronaut Bill Anders, answering his son's question about who was driving his* Apollo 8 *spacecraft to the moon*

measure. The pendulum clock was born, replacing sundials and other inaccurate clocks. I even attempted to synchronize the clocks of the world, but did not have sufficient cooperation.

N: That's odd. I thought for sure you would mention dropping those two weights from the Leaning Tower of Pisa.

G: Sorry, my friend, but that never happened. I performed an experiment in acceleration by rolling two balls down an inclined plane. Not as dramatic, I'm afraid, but sufficient to show that objects of different weight fall at the same rate of acceleration — barring wind resistance, of course.

To be honest, I would not have been famous if not for the telescope. In 1608 I learned of this Dutch invention and built one of my own. Where others turned their telescopes to the sea, I turned mine to the heavens — and I was not disappointed. The moon was not a sphere of pure crystal, as many thought, but a world of tremendous mountains and valleys. And the stars! Everywhere I looked, visible stars were multiplied a thousand fold. Clearly, the heavens were more vast than anyone had known.

Copernicus was correct. Our Earth was not the center of all of this. Jupiter had moons. Surely they revolved around Jupiter and not the Earth. I even noted dark spots on the surface of the sun.

N: The telescope was important to me, as well. Your refracting telescope was unsatisfying, however, as it often revealed inaccurate colors. I

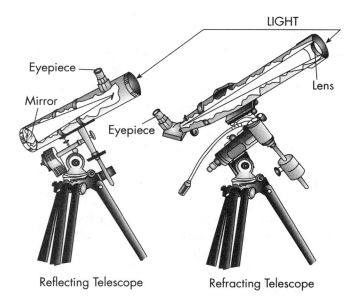

LIGHT

Eyepiece

Mirror

Eyepiece

Lens

Reflecting Telescope

Refracting Telescope

had been researching optic properties, and invented the reflecting telescope in 1668, which I found more useful even though I had the devil of a time finding accurately polished mirrors.

However, my most famous "Oh, wow" moment came much earlier.

G: When the apple fell on your head?

N: Alas, that, too, is not accurate. After graduation, I was forced to leave the university. An epidemic of plague sent me home to the country, where I had time to think. As I watched apples fall from the trees, I wondered why the moon did not fall from the sky. Kepler had spoken of gravity, but only in vague terms. I began to think of gravity extending to the moon. Kepler had shown how the periods of revolution of the planets were in direct proportion to their distances from the sun.

Mathematically, I deduced that the forces that keep the planets in their orbits must be reciprocally as the squares of their distances from the centers about which they revolve. I tested this against what was known of the moon and the Earth and found that it fit nicely. This theory of gravitation also explained the tides, as well as

the wobbling of our Earth's axis.

G: Isaac, that would have been enough to ensure undying fame. And yet you also described for the world your three laws of celestial dynamics. We finally could understand how the universe operates.

N: Yes, that was a grand moment. You would think that such a man could remember to show up for meals — could remember to go to bed for sleep. And yet these simplest of tasks eluded me. I often felt hungry and reached for food only to find it spoiled after sitting out all night. Sometimes I would wonder why I had such difficulty concentrating, only to realize that several days had passed and I had forgotten to sleep. It was often a lonely life filled with books, my friend. Why, I might never have published my theories at all if it hadn't been for the insistence of my friend Edmund Halley. And he had gained fame so easily for his prediction of the arrival of the comet that bears his name.

But I should not speak of trouble to you, my friend. You have known so much of it.

G: True, ☞ Isaac, I made a powerful enemy of the Church. After I spoke in favor of Copernicus' theories, the Church banned his books. I was no stranger to poverty, and fell to the temptation of moving to the court at

WHAT DO YOU MEAN YOU NEVER HEARD OF THE INQUISITION?

In 1478, Spanish monarchs used the name of the Catholic Church to persecute Muslims and Jews. Throughout the next 300 years, the Inquisition spread its terror to include anyone who was considered a threat to the Church or church teachings (such as Galileo). The last executions from the Spanish Inquisition came in 1781.

Tuscany — a Church stronghold. Like a fool, I continued to lecture Copernican theory. When the Inquisition banned my book, *Dialogue*, in 1632, I refused to accept what was happening.

Soon, the Inquisition brought me to Rome.

I was 70 years old when they dragged me to my knees and forced me to deny all I had said before. They threatened me with torture. On June 22, 1633, I finally gave in and declared that the Earth was, indeed, the center of the universe and that everything I knew to be true was heresy and error. Even then I was forced to spend my remaining eight years under house arrest.

 But it *does* move, Isaac. *The Earth does move!*

N: My friend, age is rarely kind. My own friend Halley even said of me, "As a man he was a failure, as a monster he was superb." In 1693 I suffered insomnia and a breakdown. For months I was unable to make any sense, even of my own book, *The Principia*. I was never the same. I even served in Parliament for one year and only spoke once — to ask someone to close a nearby window.

If I have seen farther, it is by standing on the shoulders of giants.

"That was a great conversation, wasn't it, Bud?"

"I don't know, Al. Sounded like a bit of a downer."

"Right, Bud. But we know they'll be ready when that buzzer sounds."

"They're the best, that's for sure."

"The best, Bud."

"The Dream Team, Al."

"The Dream Team, Bud."

** click **

NEWTON'S THREE LAWS OF CELESTIAL DYNAMICS

1. Every object can be described by a single quantity, MASS. And MASS possesses INERTIA. INERTIA is the tendency of a body at rest to remain at rest and of a body in motion to remain in motion.

2. In order to change INERTIA (to set a body into motion, bring a body to a stop, or change the direction or speed of a body) FORCE is required. And FORCE equals MASS times ACCELERATION (F = ma).

3. When FORCE is applied, any ACTION will result in an equal and opposite REACTION.

A C T I V I T I E S
NEWTON'S FIRST MAGIC TRICK

Did you ever see a magician pull a tablecloth out from under a fancy set of dishes? He wasn't practicing magic. He was practicing Newton's First Law of Celestial Dynamics (also known as the First Law of Motion).

A body at rest (the dishes) will remain at rest unless moved by a force (the tablecloth).

So how do you prevent the force of the tablecloth from changing the inertia of the dishes? Simple.

✔ Choose a tablecloth that is smooth and a table that is waxed, to minimize friction (a force).

✔ Pull quickly so that the force acts for only a short time.

✔ Pull straight out rather than up or down so that the force operates in only one direction.

✔ Use plastic dishes! (At least until you become an expert, and even then mail the bill to somebody else.)

NEWTON'S ACTION-REACTION

Newton's Third Law states that every action has an equal and opposite reaction. You don't need a rocket to see how this works.

Do your folks own one of those exercise machines that allows you to move as if you were cross-country skiing? Try it out.

Well, why don't you get any place when you use it?

That's right. Newton's Third Law. As you step to move forward, you are also pushing backward. The board moves back and the two forces cancel each other out.

(Wait. Do you really think it's *that* simple?)

What happens when you walk on the ground? You move forward — so what happens to the equal and opposite force?

Right! You push the ground backward in order to walk forward.

Okay, then. With all the people walking on the Earth, why doesn't their combined force change the rotation of the Earth?

If, at a pre-arranged time, everyone on Earth started walking east, could we speed up the Earth's rotation?

Why not? Hmmm...

Here at Boggler Central, we have been trying to work the bugs out of our new robot. There's been a problem with its speech programming, however, and, well, it don't talk too good. Just yesterday, it was standing in the middle of the floor hollering "Heinie toil. Heinie toil."

Nobody knew quite what it was saying, but we all thought it sounded pretty gross.

Turned out, all the poor bucket of bolts was trying to say was, "I need oil." Heinie toil — I need oil. See the problem?

So anyway, we asked old aluminum-head to come up with phrases that would be important to our five Dream Team astronomers. This is what he came up with:

1. Egg plant sorbits Annie's lips.

2. What shout think quiz is one.

3. Um, sir, vail shuns musty purse, Irv.

4. Are hurl tis apple net.

5. Ma suppose assassin her Shah.

So what did the robot say? All we know is that one phrase goes with one of the five Dream Teammates. Can you match the astronomer with the phrase? Can you figure out the phrase in the first place? Heck, can you fix our robot? This is ridiculous!

Teacher's Companion

Topics For Writing and Discussion

1. Discuss other myths. What do you know of myths other than those from Greece and Rome? Can you think of any other myths that have connections to astronomical events?

2. Do you believe in the guidelines given in daily horoscopes? How could you research whether newspaper horoscopes are accurate? (Follow the scientific method.) Can you think of any ways in which astronomical events *have* been shown to effect our behavior?

3. Discuss why our calendar is organized the way it is. Can you think of any reasons why it begins on January 1st? Why not June 21st or December 21st? Why not start the calendar at the Vernal or Autumnal Equinox?

4. Consider Kepler's Third Law. What would life on Earth be like if we were much farther from the sun (given that the sun was much more powerful, so that there would be enough solar energy for life to exist)?

5. Newton wondered why the moon didn't fall to Earth. After he determined how gravity worked, however, he realized that, in fact, the moon *was* falling to Earth. Explain how this could be.

ADDITIONAL ACTIVITIES

1. (Individual or cooperative) This historical survey is by no means comprehensive. Explore the achievements of other astronomers and form a second "Dream Team," complete with biogra-

phies. Form other "specialized" teams of five: astronomers since Newton; female astronomers; astronomers currently working; astronauts.

2. (Individual) Create a crossword puzzle of vocabulary terms (see "The Terminator" list at the start of Section 1). (Cooperative) Devise a vocabulary game and try it with the rest of the class. (All terms listed at the start of each section are defined or described in the glossary.)

3. (Multi-disciplinary):

✔ Write a myth-story connecting astronomy to unexplained phenomena.

✔ Create an imaginary interview that reveals facts about a famous astronomer and the era in which he or she lived.

✔ Choose a known constellation (or design your own from a known group of stars) and create a drawing that represents the figure or scene in your constellation.

✔ Choose an important event from the section time frame and research other world events that coincided with that event. Discuss or write about the effect each has had on the other.

4. (Challenging) Present the model shown below of Ptolemy's geocentric system. For hun-

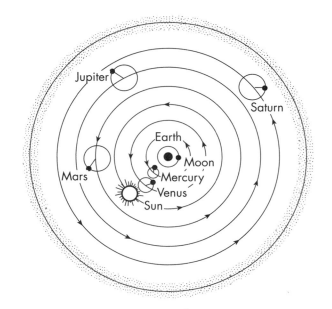

Starry sphere

dreds of years, this was the accepted version of our universe. Ask the class why it wouldn't work.

For many centuries people assumed the Earth was flat. *Prove* the Earth is not flat, and be ready to challenge your assumptions.

5. (Whole-class activity) To understand more clearly the difference between a refracting telescope and a reflecting telescope, try the following:

Refractor

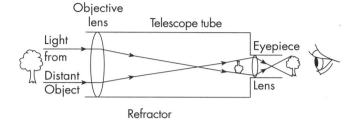

> **YOU WILL NEED:**
>
> 1 bright flashlight
>
> 2 magnifying lenses
>
> a piece of black construction paper

Cut a small arrow from the black construction paper and carefully tape it to the flashlight, so that the arrow casts a clear shadow. Turn the lights off in the room. Have one student hold a lens a few feet in front of a white wall or white poster board. Have another student shine the flashlight through the lens, moving backward until an upside-down image is formed clearly on the wall. Note how light travels through the lens to form the upside-down image. Have a third student hold the second lens up against the wall. All three should move

back together until the image is formed clearly once again, this time right-side-up. Sketch the path of light through this "refracting telescope."

Reflector

> **YOU WILL NEED:**
>
> 2 mirrors
>
> 1 magnifying lens

Darken the classroom. Line up one mirror so that it catches the light from a lighted object and reflects it to a point on the classroom wall. It will appear on the wall upside-down. Attach the second mirror to that point. Have a student

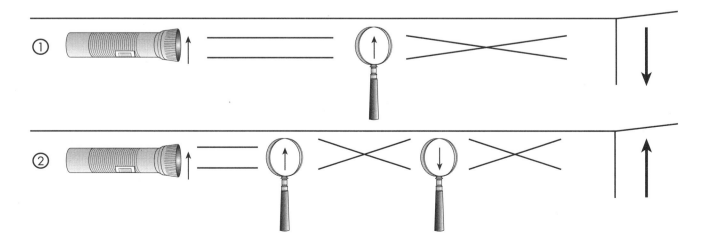

the second mirror to that point. Have a student look at the second mirror through the lens. The object will appear magnified and upright in the mirror. Illustrate and explain.

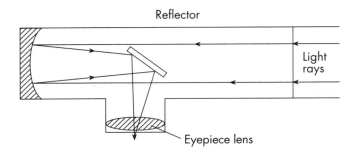

Reflector

Light rays

Eyepiece lens

	Seasons					Stars				
	Winter	Winter	Spring	Spring	Summer	Pollux	Aldebaran	Spica	Regulus	Antares
Virgo										
Gemini										
Taurus										
Scorpio										
Leo										
Pollux										
Aldebaran										
Spica										
Regulus										
Antares										

(Left axis: Zodiac for Virgo–Leo; Stars for Pollux–Antares)

The solution:

Virgo / Spica / Spring
Gemini / Pollux / Winter
Taurus / Aldebaran / Winter
Scorpius / Antares / Summer
Leo / Regulus / Spring

Bogglers' Solutions

1. Zodiac Puzzle

(page 11)

(There are several logic puzzles in this book, so it might be a good idea to take a moment and review how this sort of puzzle works.)

Display a graphing chart, as indicated in the next column, and note that the variables make up the vertices of the chart. As the clues rule out a specific combination (i.e., the first clue reveals that Regulus is not in Virgo and Aldebaran is not in Gemini), place an "X" in the box for the incorrect match. When a match is made, place an "O" in that box. Complete the chart as you complete the clues, and cross-reference as more combinations become clear.

2. Lintel and Sarcen Stones

(page 14)

No one knows the exact process used to raise the lintel stones at Stonehenge. The most widely accepted theory involves building a dirt ramp to the height of the sarcen stones, "paving" the ramp with flat stones (probably shale), and rolling the lintels to the top on log rollers. After the lintel was in place, the ramp was removed and reconstructed for the next lintel.

This would have been a monumental task, necessitating considerable societal organization. Traditionally, historians had no idea that these early semi-barbaric societies had the authoritarian structure to undertake such a massive, long-term project. The class might want to compare the construction of Stonehenge to such sites as the Great Pyramid of Giza or the Roman Colosseum — noting the differences between the societies that constructed them.

3. The Trekker-Boggler

(page 16)

This is a lesson in mathematic equations. Again, there is no answer (that I'm aware of, anyway). If someone comes up with a solution, you might offer some additional dates from any of the popular *Star Trek* reference books — to see if the formula holds up.

4. Constellation Mind-Boggler

(page 23)

The stars make a star! 5 rows, 4 stars in each row, only 10 stars total:

5. The "Boggled" Robot

(page 27)

To get the hang of this, students will need to repeat these lines slowly and out loud.

1. "A planet's orbit is an ellipse." — Kepler

2. "Watch out! The Inquisition!" — Galileo

3. "Observations must be preserved." — Ptolemy

4. "Our world is a planet." — Copernicus

5. "Mass possesses inertia." — Newton

Have the students create some phrases of their own to see if they can stump their classmates.

Section 2

It's A Beautiful Day In The Neighborhood

The Solar System

As we get ready to stretch and take a walk through our cosmic development, we're going to need to know a lot more about our neighbors. Some, after all, are pretty unhealthy places to visit. (Feeling worn down? Stop by Venus for a refreshing sulfuric acid shower!)

And we do know a lot more than we ever have before. Advanced photographic filters allow us to scan planetary neighbors for specific elements, and preliminary "visits" by *Voyager* and other spacecraft have provided so much information we are still sifting through it all — and will be for years to come.

No doubt about it. We're getting ready.

THE TERMINATOR II

Yes, the glossary is still in the back of the book.

(HE'S B-A-A-CK)

photosphere
fusion
chromosphere
corona
solar wind
sunspot
solar prominence
solar flare
isotope
terrestrial
geothermal
magnetosphere

So terminate this, already!

I'VE GOT THE SUN IN THE MORNING

I.

The Sun's Composition

The sun is made of the same things everything else is made of. Most of the elements found on Earth have already been discovered in the sun.

What does that mean? Well, it means that our Earth, other planets, and even all of us, are (as astronomer and author Carl Sagan would say) made of "star stuff."

The sun is 1.4 million kilometers in diameter and accounts for 99 percent of the entire mass of our solar system. It has what might be called a surface, what might be called a core, and what might

HOW FAR IS THE SUN FROM EARTH?

Everyone memorizes this famous distance as 93 million miles, but it's time to take a stand for individuality! Next time you want to impress someone, try this: The exact mean distance from Earth to sun is 92,750,679 miles, 2,112 feet...give or take an inch or two.

Too dorky? Okay, try the metric system for a change: 149,597,870 kilometers. Or this: Driving at 55 mph, it's a 193-year trip. What? No driver's license? Well, then, for a walk (at 7 kilometers per hour) you had better set aside 21,371,124 hours, or 890,464 days, or — um, you get a free day every 4 years, so, um — 2,438 years (by which time you could have earned your driver's license and arrived 2,200 years earlier — or whatever).

be called an atmosphere — although, since the sun is not solid, those terms are used a bit differently.

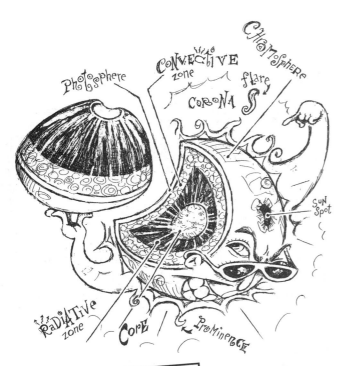

tremendous energy. The pressure at the sun's core is the weight of its entire mass, and to keep from collapsing under that pressure, the core temperature must be about 15,000,000 degrees Kelvin.

Between the sun's core and its photosphere are two areas that are defined by how they work rather than by what they are. Energy from the core's fusion furnace passes first through the RADIATIVE ZONE, by the process of (you guessed it) radiation. Energy then moves through the CONVECTIVE ZONE by the process we are familiar with when we notice that hot air rises and cold air sinks — convection currents.

As energy passes from the photosphere, it moves through three levels of the sun's atmosphere: The CHROMOSPHERE, the CORONA (the "halo" we see during a total solar eclipse), and, finally, the SOLAR WIND. And if that was all there was to it, then everything would be hunky-dory and (if there were any shade on the sun) we would have it made in the shade.

SO WHAT THE HECK IS "KELVIN"?

The Kelvin Scale is a measurement of temperature that uses the Celsius scale as its base. The "K" scale is often used in astronomy, since it begins at the coldest point possible — Absolute Zero. (At

The sun's surface is the PHOTOSPHERE, and is about 400 kilometers thick with an average temperature of 6,000 degrees Kelvin. At the sun's core, fusion converts hydrogen into helium, releasing

Absolute Zero, all molecular motion comes to a halt — sort of like study hall.) Absolute Zero is the same as −273 degrees Celsius. So you can convert "K" to "C" by subtracting 273 degrees (which won't make a whole lot of difference when you are talking about 15,000,000 degrees Kelvin).

II.
A Bit About How The Sun Works

But the way the sun works makes for some pretty wild stuff. (WARNING: The following paragraphs contain scenes of graphic violence.)

The sun rotates on its axis, much like Earth. Unlike Earth, however, it does not rotate at one uniform speed. Since the sun is not solid, it actually rotates faster at the equator (every 25.4 days) than at the poles (every 36 days). That

BUT CAN YOU DANCE TO IT?

The movement of energy to the surface causes the entire sun to vibrate, and if we could hear 16 octaves lower than we do (and if we could hear across space), we would hear a continuous hum from our mother star.

phere are darker because they are cooler than their surroundings (3,800 degrees Kelvin compared to 6,000 degrees Kelvin). Sunspots tend to form in pairs and carry opposite polarities (charges). They seem to be the result of disturbances in the magnetic fields of the sun, caused by convection and differential rotation. Sunspots can be imagined as humongous magnetic storms.

SAY WHAT?

When charged particles from the sun (ions and electrons) collide with Earth's magnetic field in our upper atmosphere, energy is released in the form of light. These lights may flash or glow in long streams. At our magnetic North Pole, the more accurate term is "aurora polaris," and when they occur near the South Pole they are called "aurora australis." Got it?

causes violent shifts of matter and energy, creating magnetic turbulences so great they even affect life on Earth.

SUN-SPOTS, recorded by the Chinese in 168 B.C., were not understood until the 20th century. These areas on the photos-

The most violent storms, however, occur in the sun's atmosphere. Because of turbulences in the photosphere,

temperatures rise in the chromosphere to about 8,500 degrees K. In the higher atmosphere, the corona, temperatures reach 1,000,000 degrees Kelvin. Here, large SOLAR PROMINENCES form like "cool" (10,000 degrees) clouds within the corona, held in place by the sun's magnetic fields. Tremendous explosions of energy, called SOLAR FLARES, push against and break free of those same magnetic fields. In a few seconds, a solar flare can release the energy of more than one billion thermonuclear explosions, expelling material out into space. No doubt about it, the corona is a violent place.

Solar flares affect Earth in several ways. The aurora borealis, or northern lights, is the visible product of these discharges.

Solar flares also affect short-wave radio communication and the natural navigation of many species of birds, and (on a bad day) can blow out transformers and interfere with satellite communications.

The solar wind is actually nothing more than the movement of these charged gases out through space. Energy, as it moves, actually does exert a force.

SAILING THE SOLAR WIND

When you flick on your flashlight, the light really does press against the glass piece in front — although the force is very weak. However, we might one day be able to "sail" space using a large Mylar sheet as a sail. A Mylar sail of 1,000 square meters could reach a speed of about 40,000 kilometers per hour after six weeks of solar wind acceleration. After all, it's one way to move through space — for free.

ACTIVITIES

MEASURING DISTANT OBJECTS

Did you ever wonder how early astronomers were able to measure the diameter of distant objects? Okay, so maybe you never wondered about that, but here's how you do it, anyway.

YOU WILL NEED:

a meter stick and a small ruler

2 3x5-inch index cards

some tape

a sunny day

a calculator (or a good brain for numbers)

Use the tape to secure the index cards to each end of the meter stick. Make sure that the cards sit up from the stick (see illustration).

Then make a pinhole through one of the cards. (Wiggle the pin around a bit to get a "clean" hole.)

Now point the meter stick toward the sun, with the pinhole end facing the sun. Aim it just right and you'll see an image of the sun projected onto the bottom index card. By the way, "aim" does not mean "look at the sun and point," okay? Scientists do NOT look directly at the sun — at least not HEALTHY scientists. Anyway, when you get the image clearly, mark off its diameter with a pen.

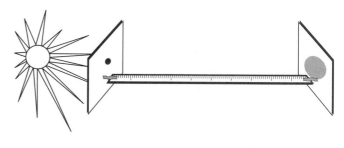

Measure the diameter of the sun's image on your card as carefully as you can. The rest is simple geometry.

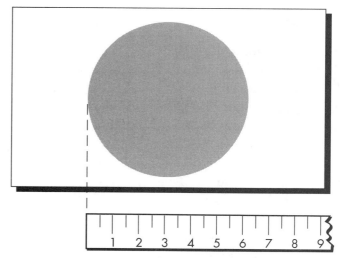

You have two isosceles triangles here (in other words, triangles that have two equal sides). One is formed by the length from the pinhole to the

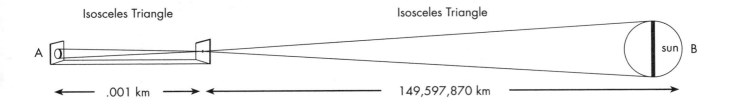

Isosceles Triangle   Isosceles Triangle

A   sun   B

←— .001 km —→ ←——— 149,597,870 km ———→

bottom card (one meter) and the width of the sun's image on that card. The other triangle is formed by the distance from the sun to your pinhole and the width of the sun itself.

We know the length of one triangle is 1 meter (0.001 kilometers) and the length of the other triangle is 149,597,870 kilometers. Set up the following formula:

$$\frac{A}{0.001} = \frac{B}{149,597,870}$$

(A = the width of the sun's image as you measured it, so you can insert the measurement and get rid of that nasty A. B = the width of the sun, which is what we're looking for.)

Okay, then, a little mathematical razzmatazz tells us that A x 149,597,870 = B x 0.001.

You can take it from there, but remember that your "A" has to be converted from centimeters to kilometers first.

(Were you expecting the answer here? Come on, let's see a little confidence — and when that runs out, check "Bogglers' Solutions" on page 73.)

USING YOUR HAND AS A SUNDIAL

QUESTION: If you were completely lost on a sunny day with nothing but your clothes (this is a family book) and a pencil, how could you tell if it's time for reruns of *Saved by the Bell*?

ANSWER: Simple. A pencil and your hand are all you need to make a sundial.

Hold your hand out flat with the palm facing up. Lay the pencil across the palm and hold the end with the thumb of that hand. As you grasp one end of the pencil with your thumb, the other end of the pencil will rise to an angle of

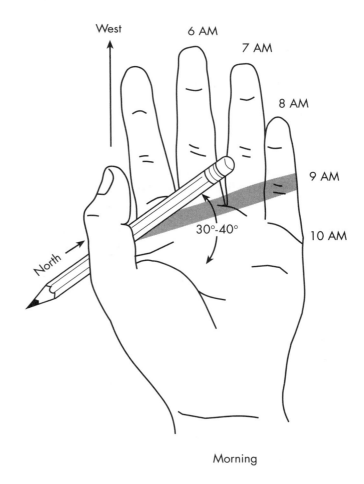

Morning

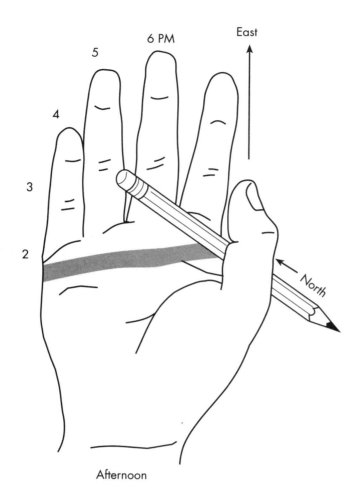

5

4

3

2

6 PM

East

North

Afternoon

about 30 or 40 degrees. This will be the gnomon of the sundial. It should point north.

When using your hand sundial in the morning, put the pencil in your left hand and point your fingers west. In the afternoon, put the pencil in your right hand and point your fingers east. The pencil should always point north. The shadow of the pencil will fall across your hand indicating the approximate time. Read the time from the tip of your longest finger around the edge of your hand. Begin at six and count one hour for each fingertip.*

---

* Wood, Robert W. *Science for Kids: 39 Easy Astronomy Experiments (3597)*. Summit, PA: TAB Books, 1991, pages 54-55. Reproduced with permission of McGraw-Hill, Inc.

... And The Moon At Night

Like the sun, we don't know everything about the moon, but we do know quite a bit. After all, 12 of us have actually walked around up there 384,400 kilometers (239,000 miles) from home on that sphere 3,476 kilometers in diameter with approximately the same surface area as Africa, although with only one-sixth of our gravity. (Notice how cleverly all that background stuff got squeezed in? That's the sign of a great book, don't you think?)

The moon is a strange place. True, there's no water (unless you count what might be frozen beneath the surface and what might exist in basins near the south lunar pole, where the sun doesn't shine) and there's no atmosphere (unless you count a layer of gases about one molecule thick). But even stranger than that is the relationship between the moon and the Earth.

IF YOU DON'T BELIEVE ME, COUNT 'EM YOURSELF!

There are 500,000 craters on the moon that are visible from Earth through powerful telescopes. However, if you counted all the craters on the moon greater than 10 centimeters in diameter, there would be more than 400 trillion (400,000,000,000,000). (So, let's see...If you counted one every second, you would be counting 24 hours a day, for 12,675,235 years. Any takers?)

For example, every once in a while, we are treated to a spectacular total solar eclipse, when the moon moves between the Earth and the sun, exactly blocking the sun. When you think about it, that's a weird phenomenon. The moon has to be the *exactly perfect* distance from Earth for such an effect. If the moon is too close, we will not see the beauty of the sun's corona. If the moon is too far, the sun's light will not be effectively blocked.

(As a matter of fact, gravitational forces are moving the moon away from Earth 3 centimeters per year. In about 40,000 years, there will be no more total solar eclipses.)

But the moon and Earth have a strange relationship in one other way: When you think about it, why do we even have a moon? Mercury and Venus have no moons, and the two moons of Mars are thought to be captured asteroids.

What was the origin of our moon? Nobody knows, but at this point, there are four theories going for the title.

1. CO-ACCRETION: Maybe the moon formed from early stellar dust and debris at the same time as Earth and the other planets. This could explain why certain elements and isotopes found on the moon match those found on Earth — and in similar proportions. However, the moon has almost no iron, and if the two were formed at the same time, iron should be much more plentiful on the moon.

2. CAPTURE: Maybe the moon is a large asteroid that was captured in the Earth's gravitational field. For this theory to work, however, the movement of the moon toward Earth as it was about to be captured would have to be so precise that the idea is not likely.

3. FISSION: Maybe, as the early molten Earth was spinning on its axis, a large bulge appeared and finally flew off into space, solidifying into the moon. If the moon was molten for a time, that could explain the loss of iron. However, the moon's present-day rotation and position do not seem to follow from such a theory.

4. IMPACT: Maybe, during the early phase of Earth's origin, an object nearly the size of Mars slammed into the Earth, sending out into space the material that would later coalesce and become the moon. This is physically believable, and would explain the similarity of elements, but the jury is still out on whether this theory could account for all that lost iron.

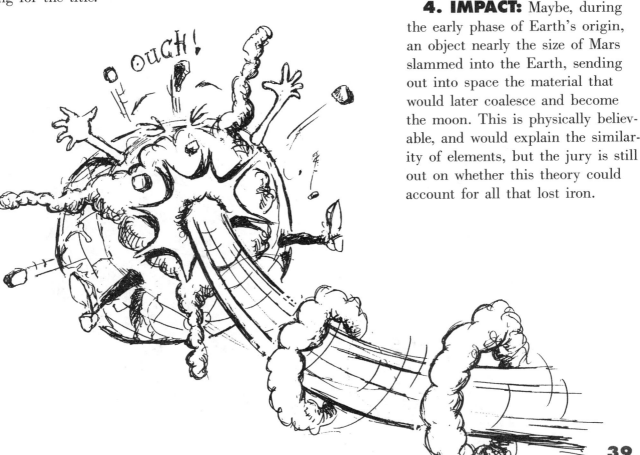

MAYBE IT ISN'T JUST THE SHOES

The moon (with some help from the sun) causes our ocean tides. The center of the ocean is pulled toward the moon by as much as 1 meter. However, the oceans are not the only things affected by tides.

During a full or new moon, gravity pulls at the land also. At high tide, the land actually rises up about 15 centimeters.

But whatever originally happened, here's what probably followed:

Heated by the constant bombardment of large asteroids and planetoids, the surface of the young moon was molten to a depth of a few hundred kilometers.

About 4.3 billion years ago, the moon's surface gradually cooled, while meteorites continued to hit the surface. Radioactive heat under the mantle forced magma up and onto the surface, creating lava seas (the maria).

The bombardment continued, churning up the rocks of the moon's surface. The last major collisions came between 3.5 and 4 billion years ago, when the Imbrium and Orientale basins

were blasted out. Since then, small meteorites have continued to hit the surface — along with the occasional larger hunk.

Hopefully, we Earthlings will one day soon go back to the moon — this time with a renewed purpose. The moon is our logical colony, and we need to gather more geological data in order to make the best of it. We've come a long way from believing the moon to be a round mirror in the sky (or a glob of green cheese), but we still have a long way to go — 384,400 kilometers, as a matter of fact.

Early Stage/Today

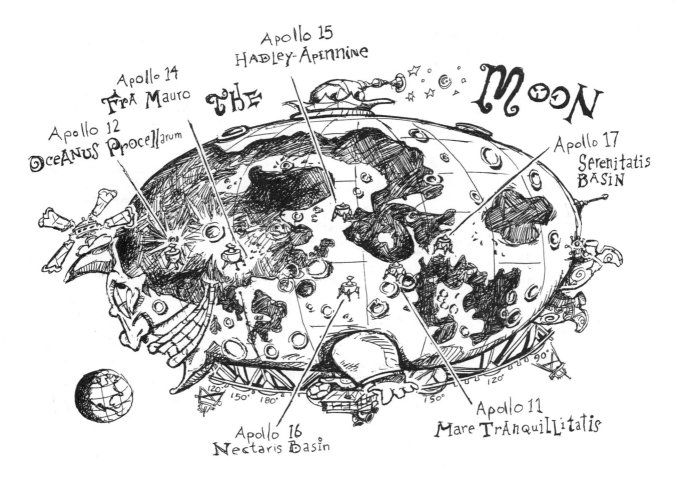

Apollo 15
HADLEY-APENNINE

Apollo 14
FRA Mauro

THE

Apollo 12
OceANUS Procellarum

MOON

Apollo 17
Serenitatis BASIN

Apollo 16
Nectaris Basin

Apollo 11
Mare TRAnquiLLitatis

ACTIVITIES

MAPPING THE LUNAR SURFACE

If you have a telescope, a spotting scope, or even a good pair of binoculars, you can observe many detailed features of the moon's surface.

Of course, you can look at the moon when it's full — and draw a map of the lunar markings. (Label the various maria and craters after you have finished drawing.) But, no matter how good an artist you are, your drawing will still be two-dimensional.

So here's the trick: To see the depth of those features, you can't look during a full moon. Wait until the moon is just past full (waning gibbous) or just before full (waxing gibbous). Then train your scope to the horizon where lunar daylight meets lunar darkness. At the edge of the shadows you can see depth. Watch for a few days in a row and you'll begin to see the moon in 3-D. (Without having to wear those cheesy movie theater glasses.)

Special Hint: The Copernicus Crater is one of the moon's best-known landmarks. The "horizon shadow" will fall across that crater on the eighth day after the new moon. Be ready!

A SNEAKY WAY TO SET A RECORD

One of the lunar samples brought back by Apollo astronauts was dated at 4.6 billion years old. It is not only the oldest moon rock known, but since it has been brought back here, it's also the oldest Earth rock known.

DEMONSTRATING THE MOON AND THE TIDES

Okay, when the gravitational force of the moon pulls on the Earth, it causes tides, and when the sun is in line to help, we have higher tides — fair enough. But why is there also a high tide on the *opposite* side of the Earth? After all, the moon isn't pulling in *that* direction.

Feeling skeptical?

YOU WILL NEED:

a round balloon

twine

graph paper

Fill the balloon with water and fight the urge to throw it at someone — at least until after the experiment. Tie the end of the balloon.

Tie the twine to the end of the balloon, and set the balloon on the graph paper.

The balloon represents the oceans of Earth. The twine represents the pull of gravity.

Mark the dimensions of the balloon on the graph paper. Now tug on the twine (quickly, but not so hard that you pull the balloon off the paper). Do you see what happens to the dimensions of the balloon? The side being pulled is drawn out (of course), but *so is the opposite side.* In astronomical terms, Earth's water is pulled by lunar gravity, and since that gravitational force is at its weakest on the opposite side, the water there actually rises against the weakened gravity. You'll also note on your graph paper that the sides of your Earth balloon experience low tides. When water is pulled on two sides, it is "borrowed" from the other sides.

Got it? Okay then, we are no longer responsible for what happens to your balloon.

It was touch-and-go for a while during the *Apollo 90210* landing. It seems that lunar astronauts Shepard Jr., Glenn Jr., Armstrong Jr., and Aldrin Jr. all had to cross the Sea of Tranquility in order to return to the lunar landing module. Unfortunately, the lunar rover could only hold two at a time.

That wasn't the only problem:

1. Armstrong Jr. and Aldrin Jr. could not be left alone together. An argument over whether two astronauts with the same last initial should have been on the same mission had gotten ugly, and each had threatened the other.

2. Shepard Jr. and Glenn Jr. both loved to reminisce about the exploits of their fathers, and if left alone, they would use up all their oxygen talking about the "old days."

The rover only holds two. How did they all arrive at the lunar module safely?

ROCK ON!

(The Terrestrial Planets: Mercury, Venus, Mars, and a Little Bit of Earth)

Mercury, Venus, Earth, and Mars...the "innies." They are family. They are terrestrial rather than gaseous planets, they are small when compared to the "outies," they possess few moons, and they share a common history of formation — to a point. (Of course, Pluto is the exception to most of these statements; but then again, Pluto is the exception to just about everything. Annoying little rock.)

YOU MEAN I HAVE UNTIL TOMORROW TO READ THAT CHAPTER?

The days on Mercury are nearly as long as the years. A Mercurial day is just less than 59 Earth days long, while a year is only 88 Earth days long. That's 1 1/2 days for each year.

Venus' calendar is even weirder. With a rotation lasting 243 Earth days and a revolution lasting 224 Earth days, the Venusian days are *longer* than the years. Then again, we've all had days that felt like that.

In order to get friendly with the terrestrial planets, you need to understand how their surfaces and their atmospheres operate.

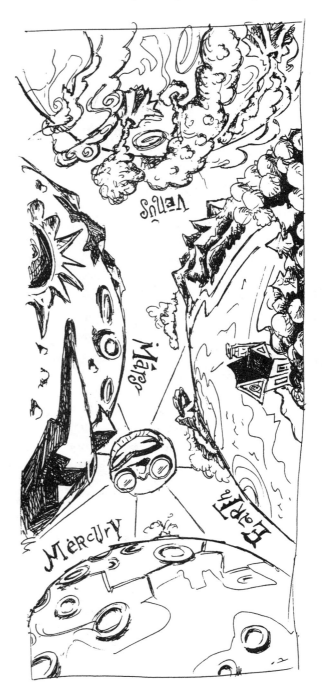

I.

The Surface

When we look at the surface of a terrestrial planet, what we see comes from a combination of three processes. IMPACT CRATERING is the term that describes the bombardment of a planet's surface by meteoroids, asteroids, and other bits of space stuff. VOLCANISM (which has nothing to do with Mr. Spock) describes the release of magma from beneath the planet's surface, and TECTONISM refers to the movement of large plates that form the crusty surface of a planet with a molten core.

As we see which of these forces is dominant, we can form conclusions about the planet.

Our moon reveals a surface dominated by impact cratering. From that we can conclude that the moon does not have a large molten core. Otherwise, there would be more evidence of volcanism. That, in turn, tells us that the moon is a cold place, since the planet does not radiate any geothermic energy. The number of craters also tells us of the absence of an atmosphere, since an atmosphere would cause the disintegration of many meteoroids.

In other words, it's dead.

Heavily cratered areas are also the oldest surfaces, since they represent the heavy bombardment associated with the last stages of planetary formation, some 4 billion years ago.

INVESTMENT OPPORTUNITIES FOR THE NEXT CENTURY!

Eighty percent of Mercury's mass is iron and nickel. Its iron/nickel core is 3,600 kilometers in diameter. At today's rate of iron production, we could mine Mercury for 650 billion years before the ore ran out.

 It turns out that Mercury is so much like our moon that astronomers often mistake photos of one for photos of the other — at least at first glance, or until they've had their morning coffee.

Mercury, however, has a more complicated history than numero-uno luna. Patterns of ridges on the surface show the movement of magma, so an extensive volcanic period must have followed Mercury's formation. The sun's gravitational tides pulling at Mercury would account for this early instability. Also, patterns of highlands tells us that Mercury does (or did?) have tectonic plates, but the ancient surface (lots of craters, remember?) tells us that those plates must have sunk down rather than "floated" into each other, as they do on Earth. All of this happened long ago, however, and by the time our moon had settled down (4 billion years ago), so had Mercury.

AH, YES, YOUR EYES — ALL PINK LIKE THE SKIES

On Earth, we enjoy pinkish sunrises and sunsets and a blue sky the rest of the day. (Of course you knew that — hang on, already!)

On Mars, things are reversed. During the day, air currents blow iron-rich dust particles from the surface of Mars into the sky — making the whole day look pink. At night, these currents are less active, the dust settles, and — badda-boom — blue skies at sunset.
(By the way, these are the same iron-rich particles that make the whole planet look red from Earth — but you probably knew that already, too.)

Mars presents a much more interesting geologic picture. The southern hemisphere is heavily cratered, which means? (— *come on, which means?* —) Right! Which means the surface is about 4 billion years old. But smoother volcanic planes, mostly in the northern hemisphere, show us that Mars had a much more active history — with volcanism that may even extend to today.

The most exciting thing about Mars' terrain, however, is the evidence that its climate was once much different than it is today. The ejected debris from craters forms patterns that suggest Mars once had a layer of water or ice on the surface. If so, what happened to it? (Hang on to that thought.)

Tectonic activity on Mars gets complicated. Most Martian tectonism seems to move vertically rather than horizontally, but the area known as the Tharsis region has several tectonic and volcanic features.

Venus is so much like Earth in size and density that it is sometimes called our twin. Yet, until recently, we knew almost nothing about it. (Well, that's family for you.) The dense cloud cover of Venus left much of the surface undiscovered until the late 1970s, when the Soviet Union obtained pictures from a series of Venera spacecraft sent to the planet's surface. At the same time, the American *Pioneer Venus* returned findings from hundreds of radar probes of the surface.

And things got interesting. Venus had a very complex geologic history, much like Earth. The few craters found so far indicate that the surface of the planet is young — probably less than 1 billion years (sure, that's old for a pint of yogurt, but it's young for a planet). Large plains indicate the spread of magma and volcanic activity. Also, the patterns of mountain ridges at the edges of those large plains hint that Venus has large tectonic plates, much like Earth.

Geologically, Venus and Earth are cosmic goom-bahs. When you get past the geology, however, the similarities end.

II.
*T*he *A*tmosphere

Mercury is hot and lifeless. Mars is cold and probably lifeless. If Venus isn't lifeless, it sure ought to be.

How did we get so lucky?

When the four terrestrial planets formed, their chemical makeup, their size, and their distance from the sun combined to send each off in a different atmospheric direction.

Mercury's early atmosphere was lost long ago. Solar radiation breaks down the chemicals of primitive atmospheres, and Mercury certainly got a hunk and a half of that. Solar winds also tend to "sweep" atmospheric gases into space. And if that wasn't enough, an early impact with a large meteor probably sent what little remained of Mercury's atmosphere scattering. Today, the thin atmosphere of the planet (a billion million times less dense than Earth's) comes mostly from the solar wind.

The atmospheres of Mars, Venus, and Earth are often compared, since Earth seems to have gotten the best deal of the bunch. All three have atmospheres rich in carbon. On Earth, however, much of that carbon has been trapped.

Trapped where? Most of Earth's carbon lies in its rock (such as limestone) and in its vegetation.

OKAY, WRAP IT UP

The surface of terrestrial planets is formed from a combination of cratering, volcanism, and plate tectonics.

Since smaller planets cool more easily, they are less likely to have deep molten cores and less likely to have floating surface plates.

Pluto is an annoying little rock.

SO WHERE DO YOU GET AN ATMOSPHERE, ANYWAY?

As the planets formed and grew, about 4.6 billion years ago, material that had collected was compressed under its own mass. Radiation and energy from impacts heated the rock, and gases were released. If the planet had a strong enough gravitational field, those gases were trapped into an atmosphere.

Or, you could just go buy one at Atmospheres-R-Us (located at the Intergalactic Mall around the corner from Rigel 5, open evenings on Wednesday).

Good thing, too, or we wouldn't be here saying to ourselves, "Good thing, too."

Venus and Mars have atmospheres thick with carbon dioxide — compared to Earth's free oxygen. Along with being unbreathable, carbon dioxide helps trap heat radiated from the planet's surface — a phenomenon we call the greenhouse effect.

These high temperatures prevent sulfur from escaping Venus' atmosphere. As a result, sulfur mixes with free oxygen to form sulfur dioxide. This then condenses into sulfuric acid rain.

Venus probably had oceans at some point in its history — although they were certainly hotter than bath water. However, once rising temperatures caused that water to evaporate, the carbon rocks were exposed to weathering — increasing the carbon dioxide in the atmosphere. Bottom line? More CO_2 means more trapped heat, which means more released CO_2, and, well, you get the picture.

If Venus had any early oceans, they were burned away. If Mars had any early oceans, they dissipated and evaporated in the thin atmosphere.

Mars is not a very dense planet, and does not have a very dense atmosphere. Therefore, the greenhouse effect accounts for only a slight change in the surface temperature — warming Mars only by about 5 degrees Celsius. Keep in mind from before, however, that Mars' surface has probably stopped evolving, so production of "new" atmospheric gases is at a minimum.

Venus seems to be just the opposite. Since it is a larger planet, Venus has a denser atmosphere than Mars. And since, unlike Earth, it's atmosphere is mostly carbon dioxide, the greenhouse effect has run wild. While the greenhouse effect has warmed Earth by about 35 degrees Celsius, Venus is heated an additional 500 degrees Celsius.

THE MOONS OF MARS

Two moons orbit Mars. Once they may have been asteroids that fell into orbit around the planet, but they are moons now. In 1610 Kepler predicted their existence. They are also described in the book *Gulliver's Travels*, which was written in 1726. Strangely enough though, they weren't actually discovered until 1877.

Mars is named for the god of war because of its red color. When Asaph Hall discovered the two moons, he named them after the mythical sons of Mars: Fear (Phobos) and Panic (Diemos). Hall then named Phobos' largest crater Stickney, after his wife. Any bets on how long that marriage lasted?

And here we are on good old Earth — with just enough carbon stored in rocks so that we don't have a carbon dioxide atmosphere; just far enough from the sun that we don't have to deal with the excess heat; just close enough that we are provided with sufficient solar energy to keep our vegetative oxygen-factory going; with just enough mass to hold it all together with an atmosphere.

Makes you wonder.

Phobos & Deimos — moons of Mars

A C T I V I T I E S

SEEING MERCURY

This is going to be tough, but you are about to see something most people never see in their entire lives.

Mercury.

Seeing Mercury is difficult because it's so close to the sun — which means that, for most of Mercury's orbit, it is either in front of or behind the big toaster. The only time Mercury is visible is when its orbit is at its "greatest eastern elongation" or "greatest western elongation" — astronomy-talk for the greatest distance to one side of the sun or the other (as viewed from Earth).

Even then, seeing Mercury isn't easy. Since it is so close to the sun, it will be visible only just after sunset or just before dawn — and will only appear near the horizon, which means you'll have to look for it despite the lights from the mall and your neighbor's oak tree, and…you get the picture.

Okay, enough whining.

So when is Mercury at its greatest elongation? There is a way to figure this out mathematically, but it's a lot easier to check a science or astronomy magazine, where such things are listed. Or just ask your science teacher — that's why he or she gets the big bucks.

Here are some hints to help you in your quest:

1. Once you have the dates of elongation, search the horizon within 30 minutes after sundown or 30 minutes before sunrise. Much more than 30 minutes and you've probably missed it. And look near the place on the horizon where the sun has set or will rise.

2. Find a place where the horizon is flat and lights are minimized. Have your folks drive you to a park someplace — heck, this is Mercury we're dealing with.

3. Mercury has phases, like our moon. You won't be able to see all the phases, however, since the more of Mercury we see, the closer it is to the sun. Viewing Mercury as a thin crescent is easiest.

4. Binoculars or low-powered telescopes are best for seeing Mercury. High-powered telescopes pick up too much light distortion near the horizon.

With a little patience and a little observing skill, you will join a select club of sky-watchers. Heck, the rumor is that even Copernicus never saw Mercury. The truth is out there.

SEEING VENUS

Quickie-quiz: What is the world's most frequently reported UFO? (No, not Elvis.)

Quickie-answer: Venus.

If Mercury is the hardest planet to see, Venus is the easiest. Tables for viewing Venus are commonly found in newspapers. Whether Venus is visible near sunset or sunrise depends on its orbital position.

You will not need binoculars to see Venus. Its dense cloud cover reflects so much sunlight that it is usually the brightest point in the sky (other than the sun or moon).

Like our moon — and Mercury — Venus also has phases. When Venus is full, however, it is farthest from the Earth and actually dimmer than when it is a crescent. The crescent Venus is sometimes so bright that, if you know where to look, you can even see it in daylight. The crescent Venus is often mistaken for the lights of an approaching plane (or a flying saucer).

So, the next time you are walking along after sundown with someone you want to impress, look casually toward Venus and say something like, "Hmm...Venus appears to be in its crescent phase, don't you think?"

We know about our three neighbor planets — Mercury, Venus, and Mars — partly from information gathered by *Viking 1*, *Mariner 10*, and *Pioneer 12* (although the order of the planets is not necessarily the same as the order of the missions). We have learned that the masses of these planets, when compared to Earth (Earth = 1), are (in no particular order) 0.8149, 0.0553, and 0.1074. We have even identified planetary features such as (you guessed it, in no particular order) Chryse Planitia, Discovery Rupes, and Ishtar Terra.

Given the following clues, can you match each planet with its mission, its mass, and its surface feature?

Sure you can.

Clue 1. *Mariner 10* never saw Chryse Planitia or anything else having to do with Mars.

Clue 2. *Viking 1* missed Venus and also the planet with Discovery Rupes and the lowest mass.

Clue 3. The first letter of only one planet matches the first letter of its mission; and you're right, Mars does not have the greatest mass.

Three clues are all you get!

TERRAFORMING INTERRUPTION

"Terra-WHAT?"

Terraforming is what you do to a place to make it like Earth (terra = earth, forming = forming).

Sure, we can set up an enclosed colony on the moon or on Mars, constructed like a land-based space station. The only problem is that to do this you need to maintain a constant atmosphere, which will keep the colony very small. Images of entire cities under a dome are pretty far-fetched. (Where would they grow their food?) A much better long-term plan would be to change the planet so that we didn't *need* a dome. Terraforming.

Terraforming

Pipeline from L.A.

Terraforming Mercury would be unlikely, since its closeness to the sun presents too many problems. Venus' atmosphere also presents problems that would be hard to deal with. The moon is close by, but how would you get it to "hang on" to an atmosphere? Nope, not a good candidate.

But Mars...

NASA has actually reviewed several theories to terraform Mars, although none has been proposed as an actual project.

True, there isn't much on Mars to make it seem "livable." But consider: Mars has water (frozen at the north pole and perhaps under the surface of the planet), it can support an atmosphere, and it has a history that seems to indicate Earth-like conditions are possible.

The main problems we would need to overcome include: a low atmospheric pressure (only about 1 percent of Earth's), a lack of oxygen (only about 0.15 percent of the atmosphere), and unbearable temperatures (a winter "high" at the north pole might be −25 degrees Celsius).

Sound undoable? Remember, these problems are all interconnected. Dealing with one will help with the others. Also, recent studies of Mars from the Hubble Space Telescope reveal that the temperature might *not* be as severe as we think, and that water might be *more* available than we think. Cool, so where do we start?

Let's say we start by spreading dark material over the northern ice cap (perhaps soot, for example). It would take a lot of soot, but we're playing for big stakes here. What would that do?

Dark colors absorb heat. This would raise the temperature of the northern ice cap, causing the frozen CO_2 to melt. Releasing carbon dioxide into the atmosphere would help trap solar radiation (remember the greenhouse effect?), further warming the planet. The CO_2 would also make the atmosphere thicker. Not only that, but any water frozen in the ice cap or under the Martian surface could be released as the temperature slowly rises.

So we raised the temperature, we unlocked some water, and we thickened the atmosphere. We still need oxygen. Hey, no problemo! It's time to transport some algae to our red sister-world. The algae will do on Mars what it does on Earth — take in carbon dioxide (of which there is plenty) and release oxygen (which we need).

Terraforming.

Don't start saving for tuition to Olympus Mons University just yet, though. Terraforming is a slow process. Even if all the pieces fell into place, it would still take thousands of years to produce an atmosphere we could survive in. Feeling let down? Don't. Even before planet-wide changes occur, there will be effects in some of the craters and canyons of Mars (where the atmosphere will "settle" first). We could establish a colony in one of those canyons.

Those first settlements are still too far in the future for us to enjoy. However, any project to terraform Mars for tomorrow will require planners, explorers, and monitors today. After all, the United States and Russia already have signed a joint agreement to cooperate in space exploration — beginning with the agreement called "Mars Together." Hmm...terraforming — sounds like a job market.

ACTIVITIES

UNDERSTANDING A TERRARIUM

A terrarium is a small example of terraforming. Examine a terrarium or find a description of one in a gardening or science book. Describe the various parts of this mini ecosystem. How does water maintain itself? How about levels of oxygen? Is there any animal life in the terrarium? How does it maintain itself? What about plants and bacteria?

It's one thing to build a terrarium, following somebody's directions. It's another thing to understand how it works. That understanding applies to any terraforming project — even one as big as a planet.

COLONIZING MARS

You have been assigned the job of organizing the first Mars settlement colony. This will need to be a self-sufficient unit, since emergency help from the orbiting space station will still be six months away.

Your job? Select the occupations of the first 100 colonists. (The Puritans established a colony in Massachusetts with 100 settlers, and only 50 survived the first winter. Even so, they managed to do all right.)

How many farmers do you want? What kinds of scientists? How many men and how many women? Any kids? Is there a doctor in the house? Be ready to back up your choices (with words, not water balloons).

THE PLANETARY TOUR

Pack Your Lunch — We May Be Awhile

Hey, groupies! Ready for the Grand Tour? It'll be wilder than the Gutbuster's last show — the one where they blew the entrails out of the first three rows of the amphitheater with a bass solo — you probably heard about that one. My name's Alfred Einstein, and I'll be your tour guide, or em-cee, or hey, whatever.

And remember, it's "Alfred" not "Albert." You might be thinking of my great-uncle Al, but that's not me. True, guiding people around the outer reaches of our solar system is a bizarre kind of job. But my re-combo band, EmCee Squared, was going nowhere, so when I was offered a gig on the Grand Tour, I assumed we'd be playing all the big halls. I thought we'd start in Madison Square Garden and finish in the L.A. Coliseum, but nooooo, we start in the Asteroid Belt and finish in the Oort Cloud.

What the heck, Uncle Al would be proud.

Anyway, we will all be entering stasis sleep as we head on out to the wide open spaces. I'll be buzzed-up just before the rest of you, so I can get your tour reports handed out.

If there aren't any questions, just put your seat back and place the stasis helmet on your head. If you can't get into stasis sleep, hit the green button. It plays classical — that always wipes me out in about a minute.

The Asteroid Belt

Buzz up, Rockers! At our tour ship's cruising speed of 50,000 kilometers per hour, we have been traveling for 198 days. And where does

UP IN THE SKY! IT'S A BIRD! IT'S A PLANE! OUCH!

Only one person was ever known to have been injured by a meteorite. On November 30, 1954, Mrs. E. Hullit Hodges was asleep in her living room when a meteorite crashed through the house, bounced around a bit, and hit her on the leg. (There are accounts, however, of an Italian monk being killed by a meteorite in 1511.)

The odds that you'll be hit by a meteorite this year are 10,000,000,000,000 to 1. In other words, you're just as likely to find Elvis' face in your chicken rice soup. *

that get us? To the inside rim of the Asteroid Belt, that's where.

But hey, if you were expecting us to dodge carefully around gazillions of little space rocks, forget it. Even here in the Belt, you can go for a long time without seeing an asteroid.

Anyhow, you all have your tour reports. Now read and enjoy the show…and wow! There's a beaut on the starboard side.

* McAleer, Neil, *The Cosmic Mind-Boggling Book.* New York, NY: Warner Communications Co., 1982, page 116.

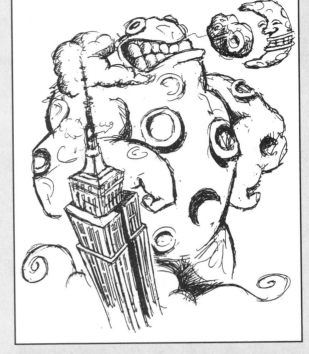

EARTH'S MOON, SCIENTISTS FIGURED IT WASN'T THE MISSING PLANET.

OTHER ASTEROID DISCOVERIES FOLLOWED IN THE NEXT FEW YEARS: PALLAS, JUNO, VESTA…. BY THE END OF THE 1800S, MORE THAN 300 ASTEROIDS HAD BEEN FOUND. TODAY, WE KNOW OF MORE THAN 4,000 BY NAME, AND THERE ARE PROBABLY HUNDREDS OF THOUSANDS MORE OUT THERE.

SOME SCIENTISTS BELIEVE THAT A PLANET ONCE DID EXIST IN THIS RANGE, BUT THAT THE GRAVITATIONAL PULL OF JUPITER EVENTUALLY RIPPED THE PLANET TO PIECES. A MORE LIKELY THEORY, HOWEVER, STATES THAT THERE WAS ENOUGH MATTER IN THIS RANGE TO FORM A PLANET (ALTHOUGH A VERY SMALL ONE). HOWEVER, DURING THE FORMATION OF OUR SOLAR SYSTEM, THE INTERFERENCE OF JUPITER'S GRAVITATIONAL FIELD DID NOT ALLOW ACCRETION OF THE PLANET. THE RESULT IS WHAT WE SEE TODAY — SMALL BITS OF ROCK, ICE, AND METAL.

THE AVERAGE ASTEROID ORBIT TAKES FIVE YEARS, BUT SOME HAVE MORE ELLIPTICAL ORBITS. SOME ASTEROIDS, CALLED APOLLO ASTEROIDS, EVEN CROSS THE ORBITAL PATH OF EARTH.

DO THEY EVER HIT US?

OF COURSE. ALL THE TIME.

TOUR REPORT: THE ASTEROID BELT

WE ARE APPROXIMATELY 330,000,000 KILOMETERS FROM THE SUN, AT THE INNER EDGE OF THE ASTEROID BELT — WHICH EXTENDS FOR THE NEXT 165,000,000 KILOMETERS. IN 1772, ASTRONOMER JOHANN BODE PUBLISHED A THEORY THAT THE GAP BETWEEN MARS AND JUPITER SHOULD CONTAIN ANOTHER PLANET — AND IN 1801, GIUSEPPI PIAZZI THOUGHT HE FOUND IT. WHAT HE FOUND, HOWEVER, WAS THE ASTEROID LATER NAMED CERES. SINCE IT WAS ONLY ABOUT ONE-THIRD THE SIZE OF THE

FORTUNATELY, ALMOST ALL OF THEM ARE SMALL. IN SPACE, THESE HUNKS OF ROCK ARE CALLED METEOROIDS, BUT WHEN THEY ENTER OUR ATMOSPHERE, THEY ARE CALLED METEORS. ANY PIECES THAT DON'T BURN UP IN OUR ATMOSPHERE ARE CALLED METEORITES, AND THE SMALL GRAINS OF "SPACE DUST" THAT FLOAT DOWN TO EARTH ARE CALLED MICROMETEORITES.

A GREAT REASON NOT TO CLEAN YOUR ROOM

Every day, about 10 tons of micrometeorites fall gently to Earth. Where do they go? Everywhere. On your lawn, on your roof, on the algebra book you left outside on the picnic table, even in your room. Some of the dust that collects on your furniture is actually from "out there."

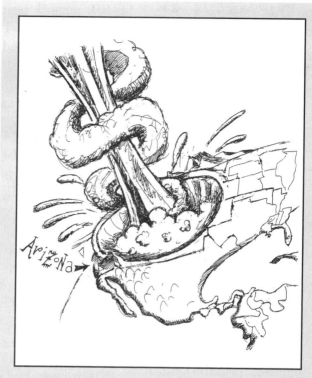

THERE ARE MORE THAN 100 KNOWN IMPACT SITES ON EARTH.

FURTHER, IT WAS PROBABLY AN ASTEROID 65 MILLION YEARS AGO THAT STRUCK THE EARTH AND LED TO THE EXTINCTION OF THE DINOSAURS (AND 90 PERCENT OF THE OTHER SPECIES ON THE PLANET AT THAT TIME, AS WELL).

METEOROIDS, METEORS, METEORITES, MICROMETEORITES — YOUR TOUR GUIDE MAY QUIZ YOU ON THESE LATER.

ONCE EVERY MILLION YEARS OR SO, A LARGER ASTEROID, PERHAPS ONE OF THE APOLLO ASTEROIDS, COLLIDES WITH EARTH. THERE ARE SEVERAL CRATERS ON OUR PLANET THAT MARK THESE EVENTS.

ABOUT 50,000 YEARS AGO, AN IRON ASTEROID 40 TO 50 METERS IN DIAMETER SLAMMED INTO THE EARTH TO FORM METEOR CRATER IN ARIZONA. THAT CRATER IS 1.2 KILOMETERS IN DIAMETER AND 200 METERS DEEP. AS RECENTLY AS JUNE 30, 1908, A METEOROID ABOUT 50 METERS IN DIAMETER EXPLODED IN THE ATMOSPHERE OVER TUNGUSKA, SIBERIA. IT DIDN'T FORM A CRATER, BUT THE FIREBALL AND SHOCK WAVE LEVELED TREES FOR MILES IN THIS (FORTUNATELY) UNINHABITED PART OF ASIA.*

PROVE IT!

Well, okay, "prove" is too strong a word, but consider this:

1. Some 65 million years ago, the fossil record changes dramatically. Nearly 90 percent of the species found in rock *before* this time cannot be found in rock *after* this time.

2. Clay deposits from this time period have been enriched with the rare metal iridium, which almost certainly came from an extraterrestrial source.

* McAleer, Neil, page 112.

Jupiter

Jupiter - the Big Gas Giant

IF IT WERE BIGGER, IT WOULD BE SMALLER

If you added up the mass of all the planets in our solar system, Jupiter would account for nearly three-quarters of it. Its diameter is approximately 143,000 kilometers. However, if Jupiter had much more mass, it would not get bigger — the extra gravitational force would cause the planet to collapse on itself. A heavier Jupiter would be smaller.

Hey, stasis-heads, this is my favorite stop. We're about five times farther away from the sun than we were on Earth. Jupiter was named after the king of the Roman gods. It is the ultimate planetary gas-ball — a place where it's easy to have your "head in the clouds" but impossible to keep your "feet on the ground." If you dig this part of the tour, sign up for the Jupiter Special — now being planned from Interplanetary Tours-R-Us.

3. If the iridium comes from an asteroid impact, it must have been a doozie, since iridium can be found in clay samples from the same time period all around the world.

4. An impact of that magnitude would have disastrous effects on the climates and ecosystems of the time. Aside from the effects of the force of the impact, dust thrown into the atmosphere would be thick enough to blot energy from the sun, creating an artificial winter and wiping out many plant species.

It isn't proof, but it looks pretty convincing.

TOUR REPORT: JUPITER

ATTENTION! DURING OUR ENCOUNTER WITH JUPITER, YOU MAY NOTICE WIDESPREAD ELECTRICAL DISTURBANCES. THESE ARE THE RESULT OF JUPITER'S TREMENDOUS MAGNETIC FIELDS. ADDITIONAL DISTURBANCES CAN BE FOUND AS WE PASS THROUGH THE BOW SHOCK REGION, WHERE JUPITER'S MAGNETIC FIELDS COLLIDE WITH THE SOLAR WIND. PLEASE REMAIN IN YOUR SEATS.

JUPITER'S MAGNETOSPHERE IS ABOUT 20 MILLION KILOMETERS IN DIAMETER, MAKING IT ONE OF THE LARGEST STRUCTURES IN THE SOLAR SYSTEM. IF WE COULD SEE JUPITER'S MAGNETOSPHERE FROM EARTH,

IT WOULD APPEAR IN THE SKY AS TWICE THE SIZE OF OUR OWN MOON.

JUPITER IS A TURBULENT MASS OF CHURNING LIQUIDS AND GASES, WITH GRAVITATIONAL AND MAGNETIC FORCES THAT RIP AND RESHAPE ITS NEAREST MOONS. MUCH OF JUPITER IS MADE OF HYDROGEN, BUT NOT IN THE GASEOUS FORM WE KNOW. ABOUT 150 KILOMETERS BELOW THE TOPS OF JUPITER'S CLOUDS, HYDROGEN IS HEATED TO NEARLY 19,000 DEGREES CELSIUS. AS IT IS HEATED, THE HYDROGEN IS ALSO COMPRESSED BY JUPITER'S GRAVITATIONAL FIELD. THE RESULT IS A LIQUID HYDROGEN, WHICH, AS IT IS HEATED AND COMPRESSED EVEN FURTHER, BECOMES A RARE FORM OF LIQUID METALLIC HYDROGEN — UNKNOWN ANYWHERE ELSE IN THE SOLAR SYSTEM.

DOES JUPITER HAVE A SOLID CORE? EVEN THOUGH THE CORE OF JUPITER IS ROCK, IT IS MOLTEN — SO THE ANSWER IS "NO."

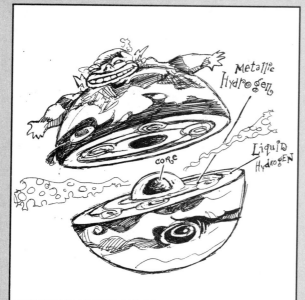

IF JUPITER SOUNDS STRANGELY LIKE OUR SUN, IT IS NO COINCIDENCE. AS IT IS, JUPITER EMITS TWICE AS MUCH ENERGY AS IT RECEIVES FROM THE SUN. IF, AT THE TIME OF OUR SOLAR SYSTEM'S FORMATION, JUPITER HAD "COLLECTED" MUCH MORE MATERIAL, ITS TREMENDOUS MASS WOULD HAVE BEGUN THE CONTRACTION AND HEATING THAT COULD HAVE EVENTUALLY LED TO NUCLEAR FUSION. BOTTOM LINE? OURS WOULD HAVE BEEN A BINARY, OR TWIN, STAR SYSTEM. CHANCES FOR LIFE ON EARTH? POOR.

JUPITER'S MOST NOTABLE FEATURE IS ITS GREAT RED SPOT, AN OBLONG BLOB OF CLOUDS 40,000 KILOMETERS WIDE. THE SPOT IS A WHIRLING STORM THAT HAS RAGED FOR AT LEAST 300 YEARS. VOYAGER DISCOVERED SEVERAL SMALLER STORMS IN THE CLOUD COVER OF THE PLANET, SOME APPEARING AS OVAL "WHITE SPOTS" NOT TOO FAR SOUTH OF "BIG RED."

IN JULY 1994, THE COMET SHOEMAKER-LEVY 9 SMASHED INTO JUPITER, ADDING TEMPORARY STORMS TO THE SOUTHERN HEMISPHERE NEAR THE GREAT RED SPOT. PHOTOGRAPHS OF THE IMPACTS OF THE COMET'S FRAGMENTS ARE MISLEADING, HOWEVER. SINCE JUPITER IS NOT SOLID, THE COMET NEVER REALLY "CRASHED." INSTEAD, AS THE FRICTION OF ENTERING JUPITER'S ATMOSPHERE CAUSED THE FRAGMENTS TO BURN, THERE CAME A POINT WHEN THEY SIMPLY EXPLODED. THERE WAS NO CRATER, OF COURSE, BUT THE EXPLOSION AFFECTED THE CLOUD PATTERNS AND REVEALED INFORMATION ABOUT THE COMPOSITION OF JUPITER'S ATMOSPHERE, AS WELL AS THE MAKE-UP OF THE COMET THAT USED TO BE CALLED SHOEMAKER-LEVY 9.

WELL...MAYBE.

ACTUALLY, NOBODY IS 100 PERCENT SURE THAT SHOEMAKER-LEVY 9 WAS A COMET AT ALL. THERE IS A CHANCE THAT THE THING THAT CRASHED INTO JUPITER WAS AN ASTEROID.

ANALYSIS OF "CRASH PHOTOS" SHOWS A LACK OF WATER. COMETS ARE MOSTLY ICE. HMM... IT WAS ASSUMED THAT SHOEMAKER-LEVY 9 WAS A COMET PARTLY BECAUSE OF THE WAY IT BROKE APART AS IT FIRST PASSED JUPITER. HOWEVER, A LOOSELY PACKED ASTEROID COULD BREAK APART AS EASILY. IT'S TRUE THAT COMETS COME FROM OUT BEYOND NEPTUNE AND ASTEROIDS COME FROM BETWEEN MARS AND JUPITER, BUT SHOEMAKER-LEVY 9 HAD BEEN ORBITING JUPITER FOR SO LONG THAT NO ONE CAN RECONSTRUCT ITS ORIGIN. ON THE OTHER HAND, THE LONG, DUSTY TAIL PRO-

Moons of Jupiter

DUCED BY SHOEMAKER-LEVY 9 WAS TYPICAL OF A COMET. A LOOSELY PACKED ASTEROID COULD PRODUCE A TAIL, BUT PROBABLY NOT SUCH A LONG ONE.

SO, DID WE SEE JUPITER GET WHOMPED BY A COMET OR AN ASTEROID? WE WILL PROBABLY NEVER KNOW. ALL OF THE EVIDENCE IS PART OF JUPITER NOW.*

Bummer, I think the copier jammed. It's probably these magnetic fields again.

Hey, no sweat! I told you this was my favorite spot. I can tell you about the moons myself. Let's see...

Jupiter has 16 known moons and a thin ring. The ring is probably formed from pieces of Metis and Adrastea, Jupiter's innermost moons, which are slowly being ripped apart by the planet's gravitational pull. The four largest moons are the Galilean moons because (say it with me, now) Galileo discovered them. These four moons offer some of the most cosmically bizarre sights in our solar system.

CALLISTO, 4,820 kilometers in diameter, seems to be one big, busted-up ice ball. You can see a series of ringed ridges in the region called Valhalla. This is probably an ancient impact site, where a large meteoroid struck the surface, melting the ice and forming — for a short time — a steaming sea. Callisto has the honor of being one of the most heavily cratered bodies in our solar system.

GANYMEDE, at 5,276 kilometers in diameter, is larger than Mercury and nearly as big as Mars. Its surface of ice is old and marked by craters, a lot like our moon. The neat thing about Ganymede is that some areas of its surface look like tile blocks. So what? Well, one theory for these strange blocks is that, while the surface of Ganymede is ice, there may be liquid water and an Earth-like crust with tectonic plates that slowly move below the ancient surface. There may be a whole underwater world there.

Does that make Ganymede like Earth? Only a

* Flamstead, Sam, "The Great Comet Crack-Up." *Discover*, January 1995, page 32.

little. Ganymede might have an Earth-like crust, but it is much less dense than our planet. That's because more than half of Ganymede is water. If you could heat it up, you would boil most of it away.

EUROPA looks like a quiet place, but don't kid yourself. That's only because it's difficult to see below its icy surface. The surface is smooth rather than cratered, with dark criss-crossing cracks. If you took a billiard ball and drew lines all over it with a pen, you'd have a pretty good model of Europa. How these cracks formed and what lies beneath are two of the hottest mysteries remaining in the Jovian family. It is possible that the cracks form as thick ice sheets, 50 kilometers or more in thickness, move slowly over the surface of Europa like massive glaciers. If Jupiter's tidal tugging creates enough internal heat, then water can exist in liquid form beneath the surface.

I know what you're thinking. Can life exist in this subsurface ocean? It's not out of the question — after all, life exists in the oceans beneath our north polar ice cap.

IO does not seem quiet at all. It's a little larger than our moon, but not nearly so peaceful. Voyager spacecraft located 10 violently active volcanoes, with a likelihood of many more that were not seen. Photographs of Io's volcanoes offer dramatic views of erupting plumes of dust and liquid sulfur. The plume from the volcano Pele rises 300 kilometers above Io and covers an area on the surface the size of Alaska.

Why such activity? Io's orbit wobbles, the effect of conflicting "tugs" from the gravitational fields of Jupiter and those of Europa and Ganymede. As it wobbles closer to Jupiter, the mother planet raises surface tides of 100 meters. Io actually heaves and crunches like a great breathing beast. No wonder it's the solar system's most volcanically active body. A great site for the next Mega-wonks concert; I mean, just think of the light show.

Picture *that* while you slip back into stasis!

Wait! Do not return to Statis until we have cleared Jupiter's Magnetic Fields!

ACTIVITIES

DEMONSTRATING CENTRIPETAL FORCE

Hey, Alfred here. Do you remember the trash-rock band, the Pulsars? No? Well, at their Lunar Base I concerts, they used to snag their old CDs in their guitar strings and fling them into orbit. It was a cute gimmick, but they overdid it.

Anyway, the secret to staying in orbit is all in how you fall.

To check this out, try this experiment.

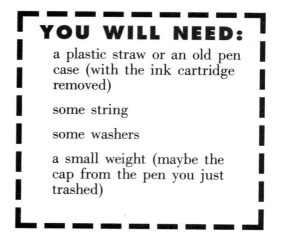

YOU WILL NEED:

a plastic straw or an old pen case (with the ink cartridge removed)

some string

some washers

a small weight (maybe the cap from the pen you just trashed)

Slip the string through the straw or pen case. Attach the weight at one end (this will be the swing end). Attach some washers to the other end (this will be the stationary end).

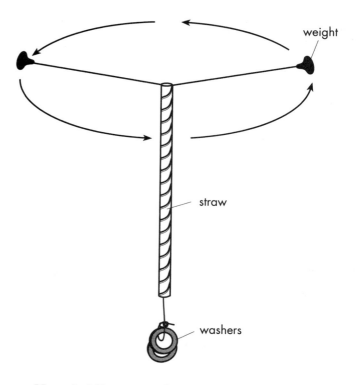

Now, holding onto the straw, let the washers dangle and swing the weight over your head. Swing just fast enough to keep the washers from rising toward the straw or falling.

(WARNING! For heaven's sake, be careful. If you hit yourself in the head, you'll look like a real dork. And by the way, don't swing it near anything breakable. Finally, if you *do* break something, make sure your folks read this warning before they write letters to the publisher.)

Okay! Looks like a dumb experiment, right? Not so fast — there are forces at work here.

Let the small weight represent a planet revolving around the sun (or a moon revolving around a planet). The washers will represent the sun (or the planet). To remain in a stationary orbit (the washers do not rise or fall), the planet must move at a certain speed.

The string represents gravity, an attraction between all bodies with mass. As gravity tugs at the moving satellite, it pulls with a force known as centripetal force. If the "planet" was not moving fast enough, centripetal force would pull it into the "sun."

Spinning the weight creates a force that, if not for the string, would cause the satellite to fly off into deep space (the living room). That force is centrifugal force.

To create a stable orbit, the speed of a satellite must be such that centrifugal force equals centripetal force. If the planet has more mass (more washers), the speed to maintain orbit would have to be faster. Try it.

Had enough? Then try looking at it this way: The speed of the satellite keeps it from falling into the planet, yet the satellite is not moving in a straight line — it's moving in a circle. That is because it's *always falling*.

An orbit happens when an object is moving so quickly that the rate at which it falls equals the curvature of the body it's revolving around. If the Earth were flat, the moon would crash into it.

Not such a simple experiment after all.

CHARTING JUPITER'S MOONS

Have you ever seen the Galilean moons of Jupiter? You *can* see them with a telescope or even a good pair of binoculars. Check in an astronomy magazine (or ask the teach) for the next time Jupiter will be visible in the night sky and where to look.

Once you find Jupiter, you will recognize its moons.

In the center of a piece of graph paper, draw a circle to represent Jupiter and mark the location of the moons that you found. Look again the next night and for as many nights as weather permits. Chart the movements of Jupiter's moons. You will even be able to see when one of them passes in front of or disappears behind Jupiter.

Okay, now. We're free from those magnetic disturbances. Back to stasis with you. Chill!

Saturn

Reveille, rockers!

You've been out for quite a while since our last stop. We've just doubled the distance from Mercury to Jupiter in order to get out here at Saturn, 1,429,000,000 kilometers from the sun. I'm required to tell you that, from this point on in our tour, there will be no communications with Earth. Any signal we send today will take 90 minutes to reach home, and an answer would take another 90 minutes — and that'll get worse from here on in. Anyway, you don't want to know the mess relativity is making with your personal lives

BUT YOU WOULD NEED A HECKUVA BATHTUB TO PROVE IT!

Saturn's density is about 70 percent that of water. In other words, the planet would float.

AND THAT GOES FOR YOUR LITTLE DOG, TOO!

Wind speeds of 120 kilometers per hour on Earth are considered hurricane force. Near Jupiter's equator, wind speeds frequently top 400 kilometers per hour. But even Jupiter's storms are a spring breeze compared to the wind speeds at Saturn's equator — which can top 1,700 kilometers per hour. A storm like this on Earth would leave nothing standing. Nada. Zilch.

back on the little blue rock. You did read the fine print, didn't you? Not to worry. We're out here in Saturnia, enigmaville — the land of the cosmic question mark.

Different ways people have seen SATURN in the past

TOUR REPORT: SATURN

SATURN, A GASEOUS PLANET LIKE ITS DISTANT NEIGHBOR, JUPITER, PRESENTS MORE MYSTERIES PER SQUARE KILOMETER THAN ANY OTHER OBJECT IN OUR SOLAR SYSTEM. LIKE JUPITER, SATURN GENERATES MORE HEAT FROM WITHIN THAN IT RECEIVES FROM THE SUN. ALL PLANETS GENERATE SOME HEAT, BUT FOR PLANETS CLOSE TO THE SUN, THAT HEAT IS DIFFICULT TO MEASURE BECAUSE IT IS MASKED BY SOLAR RADIATION. FOR THE OUTER PLANETS, HOWEVER, INTERNAL HEAT IS EASY TO MEASURE.

THE HEAT GENERATED BY SATURN AND JUPITER COMES FROM TWO SOURCES. RADIOACTIVE ISOTOPES, GATHERED AS THE PLANETS FORMED, CONTINUE TO DECAY, RELEASING ENERGY. HEAT IS ALSO RELEASED AS GRAVITY PULLS MATTER TO ITS CENTER. IN OTHER WORDS, JUPITER AND SATURN GIVE OFF HEAT BECAUSE THEY ARE SHRINKING.

WHEN GALILEO FIRST VIEWED SATURN, HE SAW WHAT LOOKED LIKE TWO "EARS" ON THE PLANET. TWO YEARS LATER WHEN HE LOOKED, THEY HAD DISAPPEARED. A WHILE LATER, THEY RETURNED.

THE "EARS" GALILEO SAW WERE SATURN'S FAMOUS RINGS. WHEN HE LOOKED LATER, HE WAS VIEWING THEM IN CROSS SECTION, AND THEY WERE NOT VISIBLE. LATER ON, THEY WERE VISIBLE ONCE AGAIN.

WHAT ARE SATURN'S RINGS, AND HOW DID THEY FORM? EVEN TODAY WE AREN'T SURE. FROM INVESTIGATIONS OF THE RINGS AROUND JUPITER, URANUS, AND NEPTUNE, AS WELL AS SATURN, WE DO KNOW SOMETHING OF HOW RINGS ARE FORMED. AS OUR SOLAR SYSTEM WAS FORMING, PLANETS AND THEIR MOONS WERE GATHERING (ACCRETING) DUST, ICE, AND DEBRIS. AS PLANETS GREW, HOWEVER, THEIR GRAVITATIONAL PULL PREVENTED CHUNKS OF MATTER NEAR THE PLANET FROM COLLECTING INTO MOONS. (THE OUTER EDGE OF THIS FIELD WHERE PLANETARY GRAVITY PREVENTS THE FORMATION OF MOONS IS CALLED THE ROCHE LIMIT.)

SATURN IS NO EXCEPTION, AND SOME OF ITS RINGS PROBABLY FORMED FROM LEFTOVER DUSTY ICE.

HOWEVER, THE MOVEMENT OF SATURN'S RINGS HINTS THAT THEY MAY BE YOUNGER THAN THAT. THEY MAY HAVE FORMED FROM COLLISIONS BETWEEN MOONS AND COMETS. OR, MAYBE THE ORBIT OF ONE OF SATURN'S EARLIER MOONS DETERI-

ORATED, AND AS IT CAME CLOSE TO THE PLANET, GRAVITY SHATTERED IT INTO PIECES.

THERE'S ONE OTHER POSSIBILITY. IT COULD BE THAT TWO OF SATURN'S EARLY MOONS COLLIDED WITH EACH OTHER, SMASHING THEMSELVES TO PIECES. THE BOTTOM LINE IS THAT, EVEN THOUGH SATURN'S RINGS HAVE BEEN KNOWN FOR HUNDREDS OF YEARS, WE STILL CAN'T TELL WHETHER THEY CAME FROM ONE OR A COMBINATION OF THESE FORCES.

THE MOONS OF SATURN ARE AS PUZZLING AS ITS RINGS. EVEN THE NUMBER OF MOONS IS IN DOUBT. ONE WIDELY ACCEPTED NUMBER IS 22, BUT THAT DEPENDS. WHEN IS A MOON SO SMALL IT SHOULD BE COUNTED AS A RING PARTICLE? AND THAT'S NOT THE ONLY QUESTION...

MIMAS (390 KILOMETERS IN DIAMETER) ORBITS SATURN EVERY 22.5 HOURS. ITS SURFACE IS MARKED BY ONE MAIN FEATURE: AN IMPACT CRATER NEARLY 130 KILOMETERS IN DIAMETER. SUCH AN IMPACT SHOULD HAVE DESTROYED THE TINY MOON. WHY DIDN'T IT?

DIONE (1,120 KILOMETERS IN DIAMETER) AND RHEA (1,530 KILOMETERS IN DIAMETER) LOOK MUCH LIKE EARTH'S MOON. HOWEVER, A LOOK TOWARD THEIR HORIZONS REVEALS A HAZY GLOW AND BRIGHT STREAKS. ARE THESE MOONS EJECTING MATERIAL FROM THEIR CORES INTO SPACE? ARE THEY COVERED WITH ERUPTING ICE-VOLCANOES?

EGG-SHAPED HYPERION (350 KILOMETERS LONG AND 200 KILOMETERS WIDE) SHOWS A RATE AND DIRECTION OF ROTATION THAT IS TOTALLY UNPREDICTABLE. ITS MOVEMENT HAS BEEN CALLED "CHAOTIC ROTATION," BUT WHAT CAUSES SUCH BIZARRE BEHAVIOR?

AND WHILE ALL THE MOONS OF SATURN REVOLVE IN ONE DIRECTION, WHAT IN THE "WORLD" IS TINY PHOEBE (220 KILOMETERS IN DIAMETER) DOING GOING IN THE OPPOSITE WAY?

THEN THERE'S TITAN...

TITAN IS THE MOST EARTH-LIKE BODY IN OUR SOLAR SYSTEM, BUT DON'T BE FOOLED BY THAT STATEMENT. IT IS LARGER THAN MERCURY AND HAS AN ATMOSPHERE MORE DENSE THAN ANY OTHER MOON. THE ATMOSPHERE IS MADE UP MOSTLY OF NITROGEN, LIKE THE EARTH'S. TITAN'S ATMOSPHERE ALSO CONTAINS ARGON, METHANE, AND OTHER GASES SIMILAR TO PRIMITIVE EARTH'S ATMOSPHERE.

IS TITAN LIKE A PRIMITIVE EARTH, AND COULD LIFE EVOLVE THERE? NOT LIKELY. TITAN'S SURFACE TEMPERATURE IS AROUND −180 DEGREES CELSIUS. ETHANE CONDENSES INTO LAKES, AND CLOUDS OF FROZEN METHANE HOVER OVERHEAD.

WELL THEN, COULD WE TERRAFORM TITAN SOME DAY IN THE FUTURE? THERE ARE SOME BENEFITS TO TERRAFORMING A PLANET WITH A "WORKING" ATMOSPHERE, AND IT WOULD HELP IF WE COULD PROVIDE A SOURCE OF HEAT. HOWEVER, MARS IS STILL A MUCH FRIENDLIER PLACE TO GO EARTH-MAKING.

BRING EXTRA SOCKS

If you were to take a hike around Saturn's outer ring (the "A" ring), covering 25 kilometers a day, the complete hike would take 95 years. A similar hike around the Earth would take a bit more than 4 years. (Of course, if hiking around a planetary body is your thing, try Mars' moon Deimos. You could do that one in 2 days!)

Stasis time, planet-jumpers.

*U*ranus *A*nd *N*eptune

LOOK AT THOSE COLORS. HAPPY PLANETS. BIG, HAPPY PLANETS.

Uranus is emerald green and Neptune is sapphire blue. Their colors have nothing to do with gemstones or anything about the planets' interiors. The colors come from methane in their atmospheres. Methane absorbs red light. Neptune reflects blue light, while Uranus (receiving more light from the sun) reflects some yellow light as well as blue.

Ouch… ginger-ale legs! One quick hint, travelers: When you go into stasis sleep don't cross your legs — man, that tingles!

I know, I know, it does-n't look like we're much of anyplace. But as a matter of fact, we are 3,690,000,000 kilometers from the sun — exactly halfway between the orbits of Uranus and Neptune. We used to stop at both planets, but with budget cut-backs — well, you know how it is.

But this is cool. Check out the port side for Uranus and out the starboard for Neptune — and no shoving. Close-ups are on the monitors and, of course, here are your tour reports.

TOUR REPORT: URANUS AND NEPTUNE

URANUS THE lazy cosmic giant

IN 1781, WILLIAM HERSCHEL BECAME THE FIRST PERSON IN RECORDED HISTORY TO DISCOVER A PLANET. AT THAT TIME, SATURN WAS BELIEVED TO BE THE MOST DISTANT PLANET, BUT HERSCHEL LOCATED A TINY DOT THAT MOVED LIKE A PLANET. IT TOOK CAREFUL WATCHING TO MAKE SURE THAT IT WAS A PLANET INSTEAD OF A COMET — BUT EVENTUALLY URANUS WAS ACCEPTED AS OUR SOLAR SYSTEM'S "MOST DISTANT MEMBER."

THE HONOR DIDN'T LAST LONG. URANUS' ORBIT WAS STRANGE. AT TIMES IT SEEMED TO SPEED UP AND AT TIMES IT SLOWED DOWN. IN 1845 JOHN C. ADAMS FIGURED OUT WHY. ANOTHER PLANET MUST BE OUT THERE, AND AS ITS ORBIT PASSED BY URANUS', ITS GRAVITATIONAL PULL AFFECTED URANUS' ORBITAL SPEED. ONE YEAR LATER, NEPTUNE WAS LOCATED, RIGHT WHERE ADAMS FIGURED IT WOULD BE.

THESE TWO PLANETS ARE VERY MUCH LIKE SATURN AND JUPITER, EXCEPT IN SIZE. SATURN AND JUPITER GREW FROM A COLLECTION OF COMETS, DUST, AND HYDROGEN GAS DURING THE FORMA-TION OF OUR SOLAR SYSTEM. URANUS AND NEPTUNE, HOWEVER, WERE OUT BEYOND THE DENS-

EST PART OF OUR SUN'S HYDROGEN ENVELOPE. THEIR SMALLER SIZE ALSO CAN BE ATTRIBUTED TO THE FACT THAT THEIR ORBITAL PERIODS ARE MUCH LONGER. THEIR FORMATION PUSHED THE RING OF COMETS OUT BEYOND PLUTO. (MORE ON THIS RING OF COMETS AT OUR LAST TOUR STOP.)

URANUS IS UNLIKE ANY OTHER PLANET IN ONE REGARD: IT IS "TIPPED OVER." URANUS' AXIS OF ROTATION POINTS ALMOST DIRECTLY AT THE SUN. THIS MEANS THAT FOR 42 EARTH YEARS (HALF OF A URANIAN YEAR), THE SUN SHINES ONLY ON ONE HEMISPHERE. FOR THE OTHER 42 EARTH YEARS, THE SITUATION REVERSES. YOU MIGHT THINK THAT THIS CAUSES WIDE TEMPERATURE VARIATIONS ON THE PLANET, BUT IT DOESN'T. URANUS' ATMOSPHERE IS AN EXCELLENT CONDUCTOR OF HEAT, AND THE PLANET MAINTAINS A STEADY TEMPERATURE.

ONE OF URANUS' FIVE MAIN MOONS, MIRANDA, IS MARKED WITH SO MANY DIFFERENT SURFACE FEATURES THAT ASTRONOMERS HAVE A HARD TIME NAMING THEM ALL. THERE ARE ANCIENT ROLLING HILLS, OVAL-SHAPED CLIFFS THAT LOOK LIKE RACETRACKS, DEEP GROOVES IN A "V" SHAPE CALLED A CHEVRON, ONE AREA NICKNAMED THE "DECK OF CARDS," AND OTHER STRANGE CANYONS, CUTS, AND CLIFFS.

SHEPHERD MOONS? WASN'T THAT A SINGING GROUP?

Shepherd moons help keep narrow rings in place. One small moon orbits just inside the ring. It moves faster than the ring of particles outside, and its gravitational pull speeds the orbit of nearby ring particles, pushing them into a higher orbit.

☞ *continued*

☞ URANUS ALSO HAS A RING SYSTEM, AND LIKE OTHER RINGED PLANETS, THE RINGS END WHERE THE ORBITS OF MOONS BEGIN. THERE ARE TWO "SHEPHERD MOONS" AT URANUS' OUTER RING, HOWEVER, KEEPING ALL THE PIECES OF DIRTY ICE IN PLACE. NEPTUNE IS THE MOST DISTANT OF THE GASEOUS PLANETS. IT IS SURPRISINGLY MORE ACTIVE THAN ITS NEARER TWIN, REVEALING SWIRLING STORMS IN ITS CLOUDY ATMOSPHERE MUCH LIKE JUPITER'S. A BIG BLUE SPOT REVEALS A TREMENDOUS, OLD STORM — JUST LIKE JUPITER'S GIANT RED SPOT. NEPTUNE ALSO HAS A FAINT RING SYSTEM AND SIX KNOWN MOONS.

☞ On the outside of the ring is another moon, moving slower than the ring particles. Its effect is just the opposite, to push particles into a lower orbit. As a result, the two tiny moons keep the ring in place.

NEPTUNE'S LARGEST MOON, TRITON, IS TILTED DRAMATICALLY FROM NEPTUNE'S EQUATOR. THIS, PLUS THE FACT THAT IT FOLLOWS A RETROGRADE ORBIT (MOVING IN THE OPPOSITE DIRECTION) AROUND NEPTUNE, SEEMS TO INDICATE THAT TRITON WAS ONCE AN INDEPENDENT PLANETESIMAL THAT MOVED TOO CLOSE TO NEPTUNE AND WAS "GRABBED" BY ITS GRAVITY. TRITON ALSO HAS THE HONOR OF BEING THE COLDEST PLACE IN THE SOLAR SYSTEM, WITH A SURFACE TEMPERATURE OF ONLY 38 DEGREES ABOVE ABSOLUTE ZERO.

ANOTHER MOON, NEREID, FOLLOWS A WIDELY ELLIPTICAL ORBIT, AND MAY ALSO HAVE BEEN CAPTURED.

ONCE-IN-A-LIFETIME CHANCE

Take note of where Neptune is right now. Since the orbital year on Neptune is 164.8 Earth years, you won't find it there again — not personally, anyway.

Thirty-eight degrees Kelvin! That is just too cool, travelers!

Now before we head off to the Kuiper Belt and that annoying little rock Pluto, a little trivia. From 1979 until 1999, Pluto was NOT the most distant planet in the solar system — Neptune was. That 20-year switch comes around every 145 years.

Now, if you've got the munchies, I'll come by with peanuts and a soda. Otherwise, head for stasis and I'll see you in about 1,300,000,000 kilometers.

and Charon, although if you check the port side viewscreen, you can see the dim haze of the Kuiper Belt. Sorry, there will be no view of Planet X, since nobody's found the thing yet. Anyway, while I'm passing out the tour reports, why don't you all stretch and switch seats — your buns have got to be getting pretty sore by now.

☞ 4.4 billion kilometers from the sun at its closest point and 7.4 billion kilometers at its farthest. If Earth's orbit were that elliptical, we would come within 19 million kilometers of Mars at our farthest point, and we would "skim" within 3 million kilometers of Venus at our closest.

Last Stop: Pluto
and the
Kuiper Belt

DON'T STICK YOUR HEAD OUTSIDE THE WINDOW, HERE COMES VENUS

Pluto's orbit is more elliptical than any other planet,
☞ *continued*

Hey, rockers! Here's where we make the big turn and head for home. We're about 6 billion kilometers out, about 40 times the distance of the Earth from the sun. Our last stop is here at Pluto

TOUR REPORT: PLUTO AND THE EDGE OF THE SOLAR SYSTEM

AFTER IRREGULAR MOVEMENTS OF URANUS LED TO THE DISCOVERY OF NEPTUNE, THE "MOVEMENT MYSTERY" HEATED UP. BOTH URANUS AND NEPTUNE SEEMED TO BE PULLED BY SOME OTHER GRAVITATIONAL SOURCE — PROBABLY ANOTHER PLANET.

IN 1930, CLYDE TOMBAUGH FOUND IT. HOWEVER, THE NEW PLANET PLUTO WASN'T MASSIVE ENOUGH TO ACCOUNT FOR THE ORBITAL VARIATIONS. EVEN THE 1978 DISCOVERY OF PLUTO'S "MOON" CHARON DIDN'T ADD ENOUGH MASS. COMBINED, PLUTO AND CHARON HAVE ONLY 1/1000 OF THE MASS NEEDED TO AFFECT THE ORBITS AS NOTED.

WAS THERE A TENTH PLANET? A PLANET X? THE SEARCH BEGAN, BUT NO ONE FOUND ANY EVIDENCE OF ANOTHER PLANET. NOT ONLY THAT, BUT MORE RECENT RECORDINGS OF THE ORBITS OF URANUS AND NEPTUNE SHOWED NO IRREGULARITIES. FURTHER, AS *PIONEER 10* AND *11* AND *VOYAGER 1* AND *2* LEFT OUR SOLAR SYSTEM, THEY SHOWED NO SIGNS OF A PULL FROM A TENTH PLANET.

THINGS DON'T LOOK GOOD FOR THE PLANET X HUNTERS. HOWEVER, IT MAY BE THAT PLANET X DOES

EXIST, BUT FOLLOWS A VERY ELONGATED ORBIT, AND IS CURRENTLY NOT IN A CLOSE ENOUGH POSITION TO AFFECT ANY PLANETS OR SPACECRAFT.

THE ROTATIONAL PERIODS OF PLUTO AND CHARON HAVE BECOME "DYNAMICALLY LOCKED." PLUTO ROTATES EVERY 6 DAYS, 9 HOURS, 17 MINUTES. CHARON REVOLVES AROUND PLUTO EVERY 6 DAYS, 9 HOURS, 17 MINUTES. THEREFORE, THE IMAGE OF CHARON IN THE SKY OVER PLUTO NEVER CHANGES — IT IS ALWAYS IN FULL VIEW, OR ALWAYS AT THE HORIZON, OR ALWAYS OUT OF SIGHT. THIS BEHAVIOR SOUNDS MORE LIKE A DOUBLE PLANET THAN A PLANET AND ITS MOON.

BOTH ARE ROCKY. CHARON HAS A WATER FROST AND PROBABLY USED TO HAVE METHANE AS WELL. HOWEVER, THE METHANE HAS PROBABLY "MOVED" TO PLUTO AND HOVERS OVER THAT PLANET IN FROZEN CLOUDS.

ASSUMING, OF COURSE, THAT YOU WANT TO CALL PLUTO A PLANET AT ALL. MANY SCIENTISTS BELIEVE PLUTO IS A LARGE COMET THAT HAS DROPPED FROM THE KUIPER BELT INTO A TEMPORARY (IN COSMIC TERMS, ANYWAY) ORBIT AROUND THE SUN.

MUCH WAS LEARNED ABOUT PLUTO AND CHARON IN THE LAST PART OF THE 20TH CENTURY, WHEN THE TWO WERE NOT ONLY AT THEIR CLOSEST TO EARTH, BUT WERE ALSO AT THEIR CLOSEST TO THE SUN (PERIHELION). THAT COMBINATION WILL NOT BE REPEATED UNTIL 2113.

WHAT ELSE IS THERE?

IF THERE IS NO PLANET X, THEN DOES OUR SOLAR SYSTEM END WITH PLUTO?

PROBABLY NOT. FAR BEYOND PLUTO PROBABLY LIES AN AREA THAT IS THE SOURCE FOR THE SHORT-TERM COMETS THAT MOVE IN TOWARD THE SUN. THIS REGION IS CALLED THE KUIPER BELT, AND ALTHOUGH IT'S DIFFICULT TO PROVE THAT IT'S OUT THERE, THE EVIDENCE IS MOUNTING.

PLUTO AND CHARON MAY HAVE COME FROM THE KUIPER BELT, AS WELL AS NEPTUNE'S MOON TRITON AND THE "ASTEROID" CHIRON.

AS URANUS AND NEPTUNE ORBITED IN A PRIMITIVE SOLAR SYSTEM, THEY WOULD HAVE "COLLECTED" MANY COMETS AND PLANETESIMALS. HOWEVER, IF THEY PASSED NEAR AN OBJECT BUT DIDN'T COLLECT IT, THEIR GRAVITATIONAL PULL WOULD HAVE ADDED SPEED TO THE OBJECT AS IT ORBITED THE SUN. ADD SPEED TO AN ORBITING OBJECT AND IT MOVES TO A HIGHER ORBIT.

OVER TIME, THE EFFECT WOULD HAVE BEEN TO PUSH A BELT OF PLANETESIMALS OUT BEYOND THE ORBIT OF NEPTUNE (AND PLUTO). THIS REGION IS THE KUIPER BELT. EVERY ONCE IN A WHILE, ONE OF THESE PLANETESIMALS DROPS FROM ORBIT AND BEGINS TO ORBIT CLOSER TO THE SUN. THESE ARE THE SHORT-TERM COMETS (COMETS THAT RETURN WITHIN 200 YEARS).

SO THEN, IS THE KUIPER BELT THE EDGE OF THE SOLAR SYSTEM? NOT IF YOU WANT TO COUNT THINGS OTHER THAN PLANETS. FOR EXAMPLE, HOW FAR DOES THE SOLAR WIND EXTEND INTO SPACE? THAT WOULD BE THE PHYSICAL REACH OF OUR SUN. THE AREA COVERED BY THE SOLAR WIND IS CALLED THE HELIOSPHERE AND EXTENDS FROM 75–125 ASTRONOMICAL UNITS (A.U.) FROM THE SUN. (REMEMBER, THE DISTANCE FROM EARTH TO THE SUN IS 1 A.U. PLUTO IS ABOUT 40 A.U. FROM THE SUN.)

AND THEN WHAT?

PERHAPS THE REAL EDGE OF THE SOLAR SYSTEM IS THE LIMIT OF THE SUN'S GRAVITATIONAL INFLUENCE. THAT WOULD BE ABOUT HALFWAY TO OUR NEAREST STAR, PROXIMA CENTAURI (4 LIGHT-YEARS AWAY). OUT THERE, AT 2-LIGHT YEARS, AT THE LIMIT OF THE SUN'S GRAVITATIONAL INFLUENCE, IS WHERE WE FIND THE OORT CLOUD. HERE, ENCLOSING OUR SOLAR SYSTEM IN A GREAT SPHERE, ARE THE REMNANTS OF ITS EARLIEST DAYS. HERE ARE THE LEFTOVERS OF PLANETARY FORMATION, THE

DUST FROM WHICH WE ALL WERE MADE. WE KNOW OF THE OORT CLOUD ONLY WHEN ONE OF ITS MEMBERS DROPS FROM ORBIT AND BECOMES A LONG-TERM COMET.

THE OORT CLOUD IS BEYOND THE CAPABILITIES OF THIS TOUR, BUT IT'S OUT THERE NONETHELESS. AND BEYOND THAT — THEN WE HAVE REALLY LEFT HOME.

the Oort Cloud

While we've been counting the big "Zs," life in the universe has been trucking along. How long? That's hard to say. The thing is, when you're moving fast through space, time does some funky stuff — relativity and all that. If my great-uncle were here he could tell you all about it. Our trip out here took about three years, and the return will take just under that. Solar gravity helps us boogie home. But time back on old Terra Firma? That's been moving more quickly. I don't know how many years will have passed when we return, but I hope nobody here left the bath running.

So, hey, it's been great having you. You're a terrific audience. I'll just warm up with a little number here, and you can hit stasis anytime you want — it won't hurt my feelings.

ACTIVITIES

ULTIMATE SOLAR SYSTEM ACTIVITY PART 1

HERE'S ONE TALL TALE!

In 1843, a comet passed near the sun. Its tail, visible across half of Earth's sky, was estimated at about 800,000,000 kilometers in length. That's longer than the distance from the sun to Jupiter.

There you have it, travelers — what a tour! We're going to take a slow turn now and start the long trek back to Earth. I'll play a couple of tunes while we're turning — that is, if I can get my axe restrung. Guitars don't go into stasis sleep, so this baby has been lying around for — well, how do you want to measure time?

YOU WILL NEED:

string

thumbtacks (or something else that will allow you to attach string to your ceiling)

a ladder (and someone to hold it)

measuring tape (we can do this in either inches or centimeters)

a pencil

a calculator

a camera (to take a photo of yourself beneath the finished product — local papers love this kind of stuff)

Okay, let's do it. Let's build the solar system. All the great science museums have solar system models — why should you be left out?

Making a model of anything is really just an exercise in scale. We know the average distances of the planets in the solar system from the sun, and we know how much room we have to work with for our model (the length of your ceiling).

Choose one end of your ceiling to be the location of the sun. If your ceiling is rectangular, choose an edge that will give the longest distance to the other side. Mark a spot for the sun, then go to the opposite wall and mark a spot for Pluto.

Measure the distance between the sun and Pluto, and convert that length to either centimeters or inches. For example, 10 feet = 120 inches, or 3 meters = 300 centimeters. Hey, you get this stuff. We'll use an example based on a ceiling that is 3 meters long. (True, it's a much "neater" number than you probably got when you measured, but the math is still the same; and geez, it's only an example for crying out loud.)

The distance from the sun to Pluto is 5,913,520,000 kilometers (3,695,950,000 miles). So, if the length of your room is 300 centimeters, then 300 centimeters = 5,913,520,000 kilometers — which means 1 centimeter = 19,710,000 kilometers. (If you want to use inches and miles, pretend for a moment that your room is 120 inches long. Then 120 inches = 3,695,950,000 miles, and 1 inch = 30,800,000 miles.)

That is your scale.

Ready? The distance from the sun to Neptune is 4,497,000,000 kilometers (2,810,625,000 miles). Set up a formula like the one below (replacing the "for example" number with the real one from your room).

$$\frac{1}{19{,}710{,}000 \text{ (ex.)}} = \frac{A}{4{,}497{,}000{,}000}$$

A little cross-multiplication and you find that

$$19{,}710{,}000 \text{ x } A = 4{,}497{,}000{,}000$$

In other words

$$A = \frac{4{,}497{,}000{,}000}{19{,}710{,}000}$$

which means

$$A = 228.15829$$

which means that the orbit of Neptune is 228.15829 centimeters from the sun on our 3-meter ceiling. (Get a reality check here and round off to 228 centimeters.) Follow the same process for miles and inches. If you came up with a distance of 91 inches, you did it correctly. So measure and make a mark already. You've earned it!

Use the distances below to measure and mark the orbits of the other planets:

Mercury	=	**57,910,000 kilometers (36,194,000 miles)**
Venus	=	**108,200,000 kilometers (67,625,000 miles)**
Earth	=	**149,600,000 kilometers (93,500,000 miles)**
Mars	=	**227,940,000 kilometers (142,463,000 miles)**
Jupiter	=	**778,330,000 kilometers (486,456,000 miles)**
Saturn	=	**1,426,980,000 kilometers (891,863,000 miles)**
Uranus	=	**2,870,990,000 kilometers (1,794,369,000 miles)**

Get that calculator smoking!

So, you've messed up your ceiling with some pencil marks. Before your mom comes in and finds the mess, you need to get some planets up there.

A WORD ABOUT LADDERS

Ladders are made to be safe. Sadly, some people aren't. If you are one of those people who trip on smooth pavement and pull a muscle jumping over a crack in the sidewalk, have someone else use the ladder. If you know you are a klutz and still can't resist using the ladder, have someone tie you down so you won't be tempted.

Ladders have two steps at the top that are not supposed to be steps. They look like steps, but they are NOT steps. They have little labels on them with big letters saying, "THIS IS NOT A STEP!"

The reason they are not supposed to be steps is because, when you get that high, your center of gravity is off-balance. The reason they still look like steps is because those who make ladders realize that most people ignore the warnings and use them anyway.

BUT DON'T! (Heck, there are warnings on lawn mowers saying not to use them to trim hedges. Oh well, as astronomers, you all know the two most common elements in the universe are hydrogen and stupidity.) If the ladder you are using doesn't have this label, it is too old. Don't use it! Okay? Are we covered?

ULTIMATE SOLAR SYSTEM ACTIVITY PART 2

YOU WILL NEED:

cardboard or heavy construction paper

(or) polystyrene (often called Styrofoam®) balls of various sizes

a calculator

the bunch of marks on your ceiling from before

a ruler

It's time to hang some planets from those marks on the ceiling, but we have a problem. Remember your scale from before? Well, unless you borrowed the ceiling of the school gym for your solar system, your scale is going to be too large. For example, our scale had 1 centimeter = 19,710,000 kilometers. Well, if the Earth is 12,756 kilometers in diameter (and it is), then our model Earth, ready for hanging is going to have to be 0.0006 centimeters in diameter! That's 6 micrometers — and you don't want to know what that would be in inches. So, unless you want to tie a string around something the size of a small bacterium and hang it from your ceiling, we need a new scale.

Okay. Go to the wall where your sun will be. Let the sun be 2 meters wide (or 6 feet). Follow the same logic as you did in computing the orbital scale above to find the size scale. The sun is 1,390,000 kilometers in diameter (868,750 miles). If 200 centimeters = 1,390,000 kilometers, then 1 centimeter = 6,950 kilometers (1 inch = 12,066 miles).

Use these diameters to calculate the diameters of your model planets:

Mercury	=	**4,878 kilometers** **(3,049 miles)**
Venus	=	**12,102 kilometers** **(7,564 miles)**
Earth	=	**12,756 kilometers** **(7,973 miles)**
Mars	=	**6,786 kilometers** **(4,241 miles)**
Jupiter	=	**142,984 kilometers** **(89,365 miles)**
Saturn	=	**120,536 kilometers** **(75,335 miles)**
Uranus	=	**51,118 kilometers** **(31,949 miles)**
Neptune	=	**49,528 kilometers** **(30,955 miles)**
Pluto	=	**2,300 kilometers** **(1,438 miles)**

Once you have completed the calculations and know the scaled-down diameters of the planets, it's time to make them. If you have access to polystyrene balls (available in craft stores), this will be easy. You will, however, need to carve those balls down to the correct size. A modeler's knife or blade works nicely.

Measure to make sure your planet is as close to the correct diameter as possible.

If you are using construction paper or cardboard rather than polystyrene, cut two circles to the correct size for each planet. Then, make one cut to the center of each circle. Use these slits to slide the two matching circles together to give a three-dimensional planet.

Attach a short length of string to each planet and, using thumbtacks or masking tape, attach the other end of the string to the correct mark on your ceiling.

Now the sun. The sun should be a strip of paper (yellow, if you have it) either 2 meters or 6 feet long, depending on the scale you chose. Attach this to the edge of your ceiling.

There you have it. A model solar system — made to correct scale.

A WORD ABOUT MODELER'S KNIVES AND BLADES

BE CAREFUL! Don't cut *toward* you; don't toss your knife to your partner; don't take it into the cafeteria; don't pretend

you are a famous sword-swallower or fencer or surgeon or mugger; and come to think of it, don't show this activity to any family lawyers, okay?

EXTRAS

1. Decorate your planets, using appropriate colors or markings.

2. Add an asteroid belt.

3. Add rings to Saturn (and Jupiter, Uranus, and Neptune, while you're at it).

4. Take a picture. Better yet, have someone else take a picture, so that you can be in it. Have somebody call the local newspaper about the young genius who constructed a solar system in his or her room! Send the article to *ODYSSEY* magazine and get ready to be famous. Turn down your first movie offer — they always come back with a better deal.

DESIGN A LIFE FORM

Design a life form. It isn't as easy as it sounds. First, choose a planet or moon that your life form will inhabit. (Remember, life forms can exist on the surface, underwater, or even in the atmosphere.)

Once you've chosen a planet, list all the facts about the planet that might have an impact on the sort of life form that could exist there. (Gravity, surface temperature, atmosphere, chemicals available, etc. — anything you can think of.)

What would your life form have to look like? What would it "eat"? Draw a picture of your life form and be ready to explain how it behaves and why.

Who knows, maybe you'll be so accurate, Earth will be able to use you as our first ambassador.

SEEING SUNSPOTS

Have you ever seen sunspots? First-hand, that is? Of course, you can't look directly at the sun — you'll damage your eyes and you won't see any sunspots anyway. Seeing sunspots is not difficult to do, if you know how to build the right equipment.

YOU WILL NEED:

a large cardboard box

a piece of heavy-duty aluminum foil

a piece of typing paper

some tape

Cut a square from one side of the cardboard box (approximately 10 centimeters square). Tape the aluminum foil on the outside of the box so that it covers this hole.

Tape the piece of typing paper on the inside of the box on the side opposite the aluminum foil.

Make a pinhole in the aluminum foil.

That's it (see "Sunspot" diagram).

(To see sunspots, you are going to have to

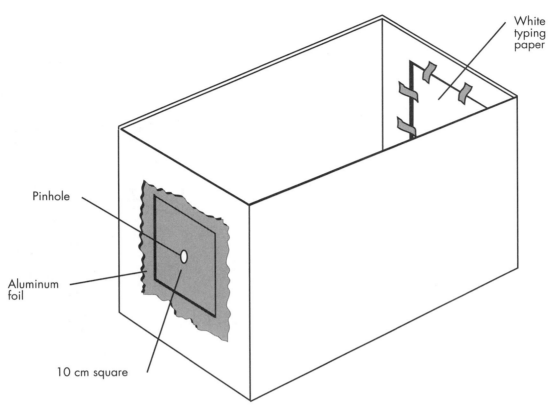

White typing paper

Pinhole

Aluminum foil

10 cm square

put your head inside this box and get it in the right position. If that sort of thing bothers your ego, then do this at night. Of course, it won't work at night, but you'll feel better.)

Point the aluminum foil side toward the sun on a cloudless day. Put your head just inside the box and maneuver the box around until the sunlight shines directly through the pinhole onto the typing paper. You now have a good image of the sun, and if there are any sunspots, you should be able to see them.

Hint 1. DON'T look directly at the sun. The spots you'll end up seeing will be from the burned-out portions of your retina. We're talking permanent damage here.

Hint 2. If the image on the typing paper is too faint, it's because too much light is getting in from the open bottom of the box. You can cover the bottom of the box, leaving just enough room for your head.

Hint 3. Don't stick your head in all the way — you'll get in the way of the image. And take off your baseball cap.

Hint 4. Sunspots can happen anytime; however, they do follow patterns. Peak sunspot times come every 11 years. The last peak was in 1991, so the next won't be until 2002. But remember, they can occur anytime.

Hint 5. If you see one sunspot, look carefully. They come in pairs (although one may be more visible than the other).

Planets have symbols. These symbols are used by scientists (and astrologers) as a sort of planetary shorthand. On the next page, is a scrambled list of those symbols. Some of them seem logical. Can you figure out the planetary matches for any of them?

Okay, okay, all right already. You want some hints:

1. Mercury's symbol is a staff that, like the one carried by the Roman mythical messenger, is entwined by snakes.

2. Venus, the Roman god of love and beauty, is represented by a hand-held mirror.

3. Earth is the home sphere — the place where it all "comes together."

4. Mars, the red planet and the god of war, is seen with a spear and shield.

5. Jupiter. What? You want a hint for everything? No way!

6. Saturn was the Roman god of the harvest, which meant a lot of heavy-duty sickle work.

7. Uranus' symbol comes from the sign for the metal platinum. (So go find out if you don't know.)

8. Neptune, the god of the sea, carries his trident.

9. Pluto was the god of the Underworld, and it'll be a cold day in the Underworld before you get a clue for this easy one.

Teacher's Companion

Topics For Writing And Discussion

1. What are the benefits of solar power as an energy source? What problems do developers of solar power have to overcome, and what strides have been made in recent years?

2. If we were able to utilize solar power in an efficient, inexpensive way, what effect might this development have on the world? Economically? Politically? Debate the advantages and disadvantages of total conversion to solar power.

3. The quest for conductivity coincides with the quest for temperatures near to Absolute Zero. Given a background in electricity, what is the relation between Absolute Zero and superconductivity? Why is it impossible to actually achieve Absolute Zero?

4. What if Earth didn't have a moon? How would life on Earth be different? Would life exist at all? (Recent research into this hypothetical situation reveals a wide range of differences — including the hypothesis that life on Earth would not have begun.)

5. How is Venus similar to Earth and how is it dissimilar? Do these comparisons have any implications for how we live and manage the global ecology of Earth?

6. Debate a plan for terraforming Mars. Consider economic and social implications as well as scientific possibilities.

7. What other plans can you think of to heat the polar ice cap of Mars as a prelude to terraforming the planet?

8. If an area of Mars was successfully prepared for settlement, what sort of social problems might settlers have to face? How could we organize the settlement so that social problems were minimized?

9. Write the job resumé of someone wishing to be considered as one of the first settlers of Mars.

10. Mining asteroids is the stuff of science fiction. Nevertheless, how might it be accomplished? We would need to bring the asteroid close to Earth and mine it while it is in orbit. Discuss futuristic plans for such an operation.

11. If Jupiter had become a star, how might its system of moons become a small "solar system"? What would be the prospects for life on any of Jupiter's moons?

12. Discuss the different ring patterns of the four gaseous planets. What are the differences in their ring systems? Brainstorm for any possible explanation for those differences.

13. Pluto is often thought of as a comet or asteroid rather than a planet. What evidence is there of this?

14. Scientists recognize that it is unlikely that our solar system has *both* a Planet X and a Kuiper Belt. Why is this such an unlikely situation? Debate the evidence first for the existence of Planet X, then for the Kuiper Belt.

15. Discuss recent discoveries that may indicate the existence of planetary systems around nearby stars. What sort of evidence might astronomers seek in this search?

16. The Voyager spacecraft have passed the orbits of Pluto and Neptune. They are on their way into "deep space," and perhaps even into the "hands" of other beings. Discuss the material on board the Voyager probes that might tell another intelligent species about us.

17. If you were sending a recording from Earth into deep space to tell "others" about our world, what would you include on that recording? What if you were limited to a 30-minute recording?

18. (As a lead-in to Section 3) In what ways is Earth "astronomically perfect" for the development of life? Orbital position? Chemical make-up? Atmospheric composition? Additionally, in what ways do plant and animal species complement each other?

ADDITIONAL ACTIVITIES

1. (Individual, cooperative, or whole-class) To note the different forms craters make in different surfaces, create a flat surface (1-inch deep) of damp sand. Create another of mud. Drop objects onto the surfaces and note the different craters formed by different-sized objects falling from different heights and onto different surfaces. How might this predict the kinds of cratering we should expect to see on various planets or moons in our solar system?

2. (Individual) Using a bright light and a small, round object (such as a quarter), close one eye and position the quarter so that it "eclipses" the light exactly. Note what happens to the eclipse as you move your quarter closer to or farther from your eye. What happens if the quarter is held in one spot, but you move your eye farther away? (Challenging) Look at a map of the path of a specific total solar eclipse over the Earth. From what you know about eclipses, determine how the sun and moon must be positioned relative to the Earth.

3. (Creative) Develop a terraforming plan for someplace other than Mars. Consider the obstacles that must be overcome and determine the order in which the terraforming project should be implemented.

4. (Individual) Create a swirling "storm" similar to Jupiter's Great Red Spot. Empty a tea bag into a jar full of water. When the tea leaves settle on the bottom, begin to stir the water. Note what happens to the tea leaves. How is it similar to what happens on Jupiter? What takes the role of the water, and what does the "stirring"?

5. (Cooperative) Combine information from the activity describing how to observe and chart Jupiter's Galilean moons. From what is known about the four Galilean moons, identify which moon is being observed. Combined charts should give a more complete orbital pattern, which can then be extended into the next month so that observers can predict what they will see.

6. (Multi-disciplinary) Chose a planet or moon and write a travel brochure. Your objective will be to put the best "public relations" face on your chosen destination. Assume there is an acceptable base on the planet (or moon). Discuss sightseeing as well as unique recreational opportunities. Illustrate your brochure. Collect and display as a futuristic travel bureau.

7. (Individual or whole-class) Add a model comet to the model solar system. Determine the shape and direction of the tail in relation to where it will be hung in the model. Construct both the ion tail and the dust tail.

8. (Multi-disciplinary) What would a futuristic, interplanetary Olympics be like? Describe extraterrestrial sports — (either how today's games might be different or new games which might be invented). High-jumping on the moon? Solar-sailing on Mercury? Try also to describe what it would be like to participate in some of these events.

9. (Cooperative review) Small groups or pairs should create an order of the planets (or an order of the moons of Jupiter), with the intention of having other students guess what the order is based upon. The planets are almost always given in order of distance from the sun. What about order of size? Get creative and attempt to stump classmates.

Bogglers' Solutions

1. Measuring Distant Objects

(page 36)

1,357,000 km

2. The Light of the Full Moon

(page 40)

When the moon is half full, the portion illuminated is very rough and mountainous and does not reflect light well. When the moon waxes full, smoother surfaces are illuminated, reflecting much more light back to Earth.

3. Lunar Rover Puzzle

(page 42)

There are several combinations that work for this puzzle. One example is as follows:

Trip 1. Scott and Armstrong cross, Scott returns.

Trip 2. Scott and Aldrin cross, Aldrin returns.

Trip 3. Aldrin and Glenn cross.

4. Planetary Mission Match-Up

(page 48)

A logic puzzle, employing one more variable than the logic puzzle in Section 1. Solution chart and matches are as follows:

Mercury	*Mariner 10*	0.0553	Discovery Rupes
Venus	*Pioneer 12*	0.8149	Ishtar Terra
Mars	*Viking 1*	0.1074	Chryse Planitia

Discovery Rupes is a large "crack" in Mercury's surface, evidence of how the planet contracted as it cooled.

Ishtar Terra is a continent the size of Australia in the northern hemisphere of Venus.

Chryse Planitia was the landing site of *Viking 1*.

5. Planetary Symbols

(page 69)

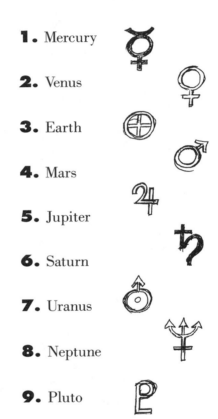

1. Mercury

2. Venus

3. Earth

4. Mars

5. Jupiter

6. Saturn

7. Uranus

8. Neptune

9. Pluto

SECTION 3

OPERATIONS MANUAL

Spaceship Prototype OMICRON IV

Please read all directions before initiating operations. Owner negligence is considered a violation of warranty.

The spaceship prototype OMICRON IV, often referred to as Earth, was constructed for your living comfort. With proper care and maintenance, your Earth will deliver billions of years of flexible living as you journey interstellar space. The enclosed manual will offer vital tips for the proper care of your spaceship. For more complete instructions, we recommend you reference *Recorded History, Consciousness, and Scientific Development*. Once again, thank you for your confidence in the OMICRON IV model. Enjoy Earth.

The Management

WARRANTY

Of all the bodies in the solar system, we know our Earth best. We have studied our oceans, our deserts, our polar caps, and the layers of our atmosphere. We have photographed our planet as a blue sphere in space, and we have photographed the smallest of the organisms that inhabit it. We recognize the spectacular coincidences of climates and chemicals that made all of this possible, and we recognize that, for the foreseeable future, this is the only home we have.

> **"We shall not cease from exploration. And the end of all our exploring will be to arrive where we started and know the place for the first time."**
>
> *T.S. Eliot*

Why, then, does it seem we act so often out of ignorance? We are, after all, only one species. There are 1,700,000 species of plants and animals we know of, and another 10 to 30,000,000 we don't know of. And we are the smallest group. There are only about 5,000,000,000 of us, while there are 580,000,000,000,000 water animals. There are 92,500,000,000,000,000,000,-000,000,000 worms and termites on this same planet, and at least 2,000,000,000,000,000,000,-000,000,000,000,000 bacteria!*

Yet, here we are — the dominant species on a small terrestrial planet orbiting an ordinary yellow star in one arm of a galaxy from a cluster of galaxies in one small segment of the universe. Each of us is allowed to ride along on the outer skin of this planet for about seven dozen trips around that yellow star.

And, like it or not, each of us is responsible for the care and maintenance of Spaceship Earth.

In the last part of the 20th century, questions regarding maintenance have involved difficult, global problems. Are we adding to our planet's greenhouse effect? Are we endangering the protective layer of ozone in our atmosphere?

* Moeschl, Richard, *Exploring the Sky*. Chicago, IL: Chicago Review Press, 1993, page 149.

Are we destroying the biodiversity of our planet, interrupting the food web in ways we can't even imagine?

Earth is a spaceship. We are enclosed within the atmosphere of our ship. The mechanisms of our ship provide a source of food, water, air, and energy. And like any self-contained spaceship, if we damage those mechanisms, we have no choice but to face the consequences.

Spaceship Earth comes with a full warranty. The parts are in good working order and are sufficient to sustain us. If you examine the fine print on any warranty, however, you'll see that there is no guarantee of service if the product is mishandled or abused.

SPECIFICATIONS

I.
*I*nterior

Earth's history goes back more than 4 billion years, but most of the Earth's surface is less than 100 million years old. Earth is constantly resurfacing itself and recycling its crust.

For at least part of the time of Earth's formation, it was molten (the energy of continuous bombardments during the solar system's formation would see to that). At that time, the liquid materials separated, with the denser "heavy" metals (mostly nickel and iron) sinking toward the center. Intense pressure compacted these metals to form a solid INNER CORE. Surrounding that solid core is a molten OUTER CORE, made up mostly of iron.

Between the core and the crust is the MANTLE, mostly molten basalt.

Above the mantle exist three "surface" layers. The third (lowest) layer is the source of the magma we see bubbling up (or exploding) out of volcanoes. The second layer is a mushy mineral mix called the SHALLOW MANTLE. The first (top) layer is the CRUST or LITHOSPHERE, or, as we like to say, home.

And it's a pretty thin crust, too. While the inner core is 1,200 kilometers thick, and the outer core is 2,200 kilometers thick, and the mantle is 2,000 kilometers thick, the terrestrial crust is a mere 20 kilometers thick, on average.

DO NOT ATTEMPT TO PROVE THIS AT HOME

If the Earth was a 1-meter-deep bucket of pudding, and we left it on the shelf overnight, the skin that would form would be about the right proportional thickness of our Earth's crust.

Too gross? Okay, then, if you are 5 feet tall and you let your height equal the radius of the Earth, the Earth's crust would be only one-fifth of an inch (45 millimeters) thick — about the thickness of a peanut butter and jelly sandwich on white bread after your best friend Chub sits on your lunch bag all the way to school. Or, if you covered yourself with 3 feet of vanilla pudding...oh, forget it.

THE TERMINATOR

ATTENTION! Before you read your "Operations Manual," you should familiarize yourself with the following operational terms:

Thank you. Continue to enjoy your Earth. Have a nice day.

magma
subduction
Pangaea
greenhouse effect
convex
concave
chromatic aberration
spectroscopy
microgravity
rehydration

The molten iron in Earth's outer core circulates at about 3/4 of a meter per hour. This motion sets up electrical currents establishing Earth's magnetic field, in case you were wondering where that came from.

ACTIVITY

DEMONSTRATING EARTH'S MAGNETOSPHERE

Earth's magnetosphere protects us from the harmful radiation of the solar wind. How? Glad you asked.

YOU WILL NEED:

a good magnet (something smaller than the ones used to pick up junked cars but larger than the ones on your refrigerator)

iron filings
a sheet of paper towel

Place the magnet in the center of a table, and lay the paper towel over the magnet. The magnet is going to represent Earth's magnetosphere. Picture Earth within the boundaries of the magnet, just as it is within the boundaries of our magnetosphere.

Holding a pinch of iron filings about two feet over the paper towel, slowly sprinkle the filings down on the towel. The iron filings will represent the solar wind that, like the filings, contains ions that react to magnetic fields.

Observe the pattern of filings on the paper towel.

So what?

When charged particles from the solar wind come within the range of Earth's magnetosphere, they are deflected, just as the filings were deflected from the center of the magnet. Radiation from the solar wind would be harmful to us, so the magnetosphere protects us.

Not bad. Now clean up the filings before the dog gets into them.

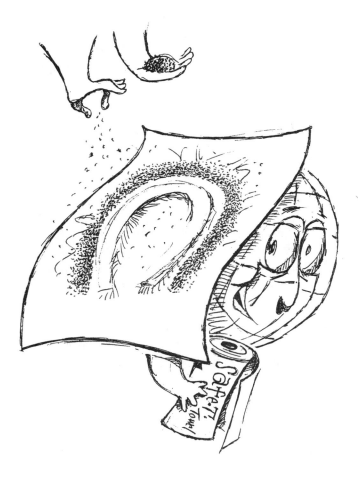

II.

Surface

Whether above or below sea level, the Earth's crust covers the molten mantle underneath. The fact that only 30 percent of the lithosphere is visible as dry land doesn't mean the crust isn't everywhere.

It's good for us that the Earth's interior is completely covered (with the occasional volcano as an exception), but there is one problem — a problem for the Earth, anyway.

How do you get rid of internal heat when you're surrounded by the lithosphere?

Our Earth is constantly generating internal heat. At this point in planetary history, most of that heat comes through the decay of radioactive materials. Heat must be dissipated some-

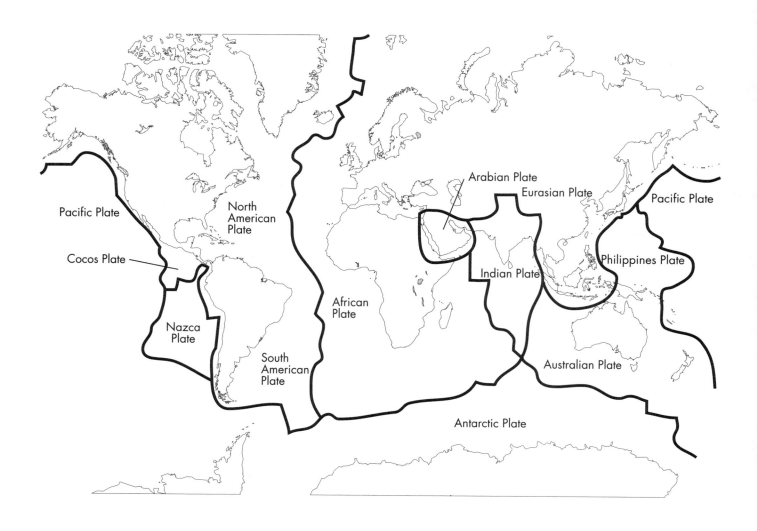

how. Volcanoes dissipate heat, but to depend on volcanism for heat dissipation would mean a lot more volcanoes around than we probably want to think about. Volcanism works on Io, but you wouldn't want to live on Io.

Heat also can be dispersed through conduction, slowly drifting to the surface much the way soup eventually cools in a bowl. Conduction, however, is not very efficient, and only works well on smaller planets, such as Mercury or Mars.

Then there's plate tectonics.

The Earth's lithosphere isn't really one, spherical crust. It is made up of eight large plates and about 20 smaller ones, which fit together and float around on top of the magma. "Tectonics" refers to various kinds of construction, so the construction of the Earth's lithosphere from floating plates is plate tectonics. Pretty logical.

Like a swimming pool filled with bathers floating on air mattresses, these plates bump and push each other over the magma. Being much more massive than your average air mattress, however, when these plates bump into each other, things happen.

Things like volcanoes and earthquakes and even mountains.

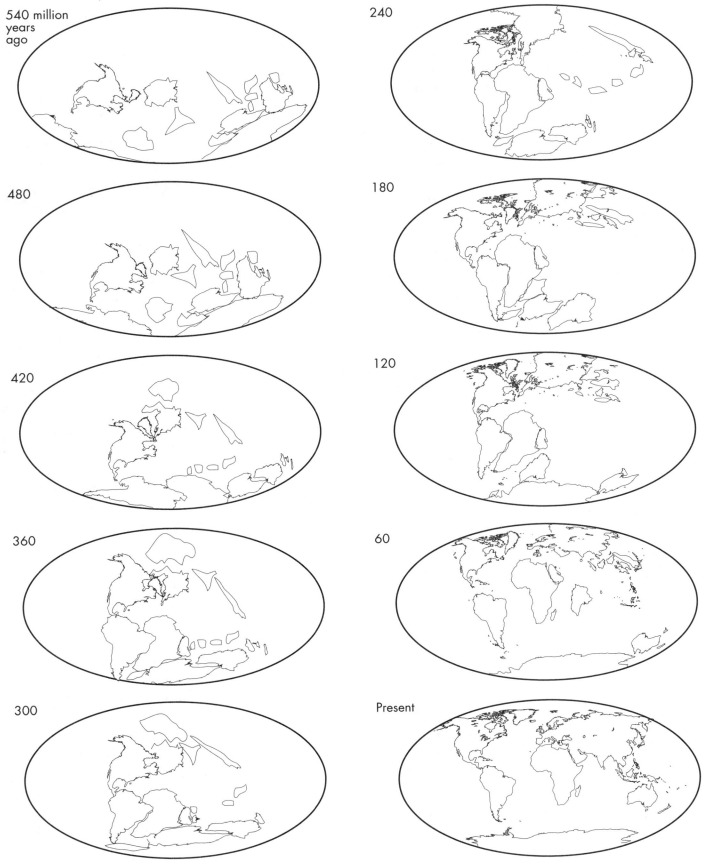

540 million years ago

480

420

360

300

240

180

120

60

Present

The North American Plate, for example, floats westward into the Pacific Plate at the rate of about 5 centimeters per year. When that happens, the continental crust is pushed up on top of the oceanic crust. The continental crust "wrinkles" as it is lifted, and mountains form (slowly, of course). The Sierra Madres are examples of this sort of mountain forming. Meanwhile, the Pacific Plate is pushed back into the magma ("subducted" is the term, in case you're really getting into this), where it eventually becomes molten. Subduction of the crust is like sticking a cold spoon into hot soup. The spoon heats up, but the soup also cools a bit. Subduction, then, is one way to dissipate heat through plate tectonics.

Also, the fault lines where the two plates rub together are sources of geologic violence, with earthquakes and volcanoes the main symptoms. (The San Andreas Fault is probably the most famous North American example of this.)

That means more heat dissipation.

But wait a minute! If subduction destroys some of the lithosphere, over a period of millions of years we would have some serious holes in our crust. Where does the new lithosphere come from?

Look eastward.

As the North American Plate moves west, it is moving away from the Eurasian Plate. That means that, somewhere under the Atlantic Ocean, two plates are separating. When that happens, magma is released to form new crust. The Mid-Atlantic Ridge is an example of new crust. As magma cools to form new crust, there is major-league heat dissipation.

As Earth's oceanic plates recycle themselves, our planet can disperse its internal heat.

Because Earth has a molten interior, centripetal force causes it to bulge as it spins. Theoretically, you are closer to the center of the Earth if you are standing at the North Pole than if you are standing at the equator — but you might be too cold to appreciate it.

By the way...Earth's surface rotation also causes the Coriolis effect. As air travels from cold polar regions to the warm equator, it is turned by the Earth's spin. So is everything else that isn't nailed down. Everything in the northern hemisphere is turned to the right. Everything in the southern hemisphere is turned to the left.

Twisting weather systems (tornadoes, hurricanes, typhoons) exhibit the Coriolis effect as well.

YES, THE RUMORS ARE TRUE!

Water going down a drain in the northern hemisphere turns in a clockwise spiral. South of the equator, the spiral is counterclockwise. Of course, your bathtub has to be level for the Coriolis effect to work.

And at the Equator? Nothing. The water may spiral, but the direction is not predictable. It just goes down. If it doesn't, don't call a physicist, call a plumber.

III.

Atmosphere

Spaceship Earth does not end with the lithosphere. Life support systems are also contained within the atmosphere.

The atmosphere itself is divided into six layers, beginning with the TROPOSPHERE. This lowest level contains 90 percent of the mass of the entire atmosphere, as well as the clouds, storms, rain, dust, smoke, and winds that affect

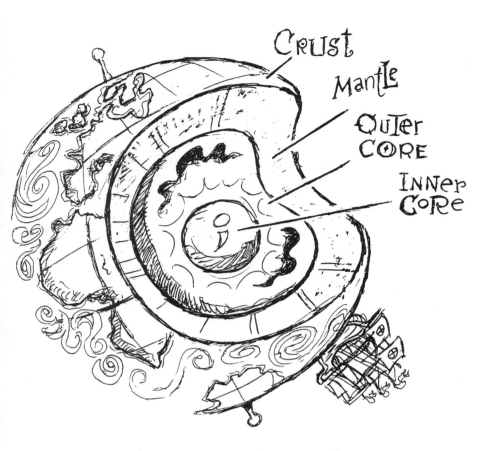

CRUST
Mantle
Outer CORE
INNer CORE

layer, the helium layer, and the hydrogen layer; but you've probably just about had it with layers.)

The IONOSPHERE contains (you guessed it) ions (electrically charged particles) that not only reflect radio waves (that's why radio signals can "bounce" over great distances on Earth), but also intercept charges from solar flares — giving the rest of us an occasional show of aurorae.

The outermost part of our atmosphere is the EXOSPHERE. Here is the last trace of gases trapped by Earth's gravity. Beyond the exosphere is deep space.

our weather. The troposphere is thinnest at the poles (8 kilometers) and thickest at the equator (18 kilometers) and contains the mix of 78 percent nitrogen, 21 percent oxygen, and 1 percent argon that we all refer to as "air."

Extending 40 kilometers above the troposphere is the STRATOSPHERE. The stratosphere contains the jet stream — winds that blow from west to east at about 450 kilometers per hour. The jet stream moves just below the cold polar air mass, so that it passes over the United States during winter months and over Canada during summer months.

If the stratosphere is home to some wild wind currents, the MESOSPHERE, also 40 kilometers high, is home to some crazy temperature variations. The lower level of the mesosphere may be +77 degrees Celsius, while upper levels drop to −101 degrees Celsius.

The next layer of our atmosphere is the THERMOSPHERE, so named because solar energy builds up here. (The thermosphere also is divided into the nitrogen layer, the oxygen

WELL, SORT OF...

The exosphere is not really the outer boundary of the Earth's atmosphere. Earth's magnetic field spreads out about 6,500 kilometers into space. The MAGNETOSPHERE is home to two radiation fields called the Van Allen Belts, after their discoverer James Belts (only kidding — it was James Van Allen).

A C T I V I T Y

DEMONSTRATING THE CORIOLIS EFFECT

Does the Coriolis effect sound weird? Actually, it's pretty easy to see how it works.

YOU WILL NEED:

a globe

a piece of chalk

You know already that the Earth rotates in an easterly direction (looking down from the North Pole, this is counterclockwise). It's also true that air moves (by convection) from colder climates toward warmer climates. That means air from the poles tends to move toward the equator.

So try it. As you rotate the globe, use the chalk to draw a line from the North Pole to the equator. (If you have a hard time working both movements at once, get a friend to rotate the globe — then try to pat your head with one hand while rubbing your tummy with the other.)

So what does the chalk line look like?

You can imagine that, as winds continue to move in this pattern, spiraling wind systems (storms) form in opposite directions in the two hemispheres. Storms in the southern hemisphere move in a counterclockwise direction, and move clockwise in the northern hemisphere. (Want to make your teacher crazy? Ask what happens if the storm crosses the equator.)

IV.

Development

If we look at the history of Spaceship Earth, we talk in millions and billions of years. For a different look at Earth's history, however, we can set the whole thing up on a 24-hour scale instead.

12:00:01 a.m. — It is Day 1. The construction of Spaceship Earth begins some 4.6 billion years ago.

1:00 a.m. — (4.4 billion years ago) A large body has collided with the still-molten Earth, throwing off a tremendous mass that collects to form the moon.

1:02 a.m. — The dust cloud that has completely hidden the sun from view dissipates.

3:07 a.m. — (4 billion years ago) The frequent bombardment has slowed, and Earth can form a stable crust.

5:12 a.m. — (3.6 billion years ago) Life may have come and gone before this, but by now it is here to stay.

1:30 p.m. — (1.9 billion years ago) Plant species have been working on the already-changing atmosphere, which has now become primarily nitrogen/oxygen rather than carbon dioxide.

8:00 p.m. — (690 million years ago) The conditions have been right for animals, but our first record of animal fossils doesn't appear until now.

10:35 p.m. — (270 million years ago) The continents of Earth form one massive land mass, called Pangaea. Within 7 minutes, Pangaea shatters and the continents begin to drift.

11:00 p.m. — (197 million years ago) Dinosaurs are seen in significant numbers. They will reign until 11:40 p.m. and suddenly disappear.

11:59 p.m. (11:58.5 p.m. if you want to stretch a point) — The first of our direct ancestors walks the Earth around 4 million years ago.

11:59:59.8 p.m. — 10,000 years of civilization fits into one-fifth of the last second on the clock.

12:00 — The present.

12:01 — ?

A C T I V I T I E S

DEMONSTRATING PLATE TECTONICS — MAYBE

WARNING! This is going to sound dumb, but it isn't, so do it anyway.

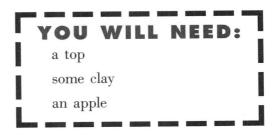

YOU WILL NEED:
a top
some clay
an apple

Spin the top. (This may require some practice, since some people are naturally spintifically challenged. Nevertheless, you need to be able to spin the thing successfully, so either practice or get a partner who isn't spin-deficient.)

Now attach a small bit of clay to the side and spin it again.

Observe.

The top spins fine when its center of gravity is unchanging. However, as soon as you upset the balance of mass (by adding a piece of clay to one side), the top spins differently — or not at all.

What do you think happens to Earth's rotation on its axis as continents float about to new positions on the globe (plate tectonics)?

Take the apple, sit back, and eat it slowly while you think.

Yes. The Earth's rotation *has* shifted in the past. In the past 60 million years, it has shifted about 8 degrees (about 20 degrees if you go back 200 million years). These shifts certainly caused tremendous changes in Earth's geological history.

MODEL OF PLATE TECTONICS — MAYBE

YOU WILL NEED:

a plastic sphere, or polystyrene sphere covered with a layer of papier-mâché (either should be readily available in a hobby or craft store)

paints or colored markers

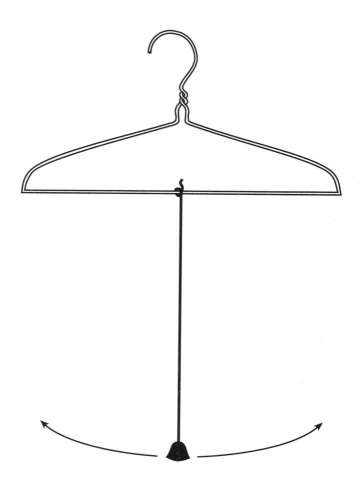

Refer to the chart on page 79 illustrating the position of the continental plates at times in Earth's past.

Choose one era, and make a model globe based on the indicated plate positions.

If classmates make other models from other eras, the entire collection will represent a tectonic plate history of Spaceship Earth.

Sure, you know the world rotates, but how do you prove it? Here's a method developed by J.B.L. Foucault in the mid 1800s:

YOU WILL NEED:

a wire coat hanger

some string

a weight

Tie the string to the center of the bottom of the wire coat hanger. Tie the weight to the other end of the string.

Now, swing the weight gently so that it swings in the same direction as the two shoulder prongs on the coat hanger. Slowly turn the coat hanger so that the shoulder prongs are at right angles to the swinging weight.

What happens to the direction of the swinging weight?

Correctamundo. It doesn't change direction. So what? So plenty!

Turn the experiment around in your imagination. What if *you* were riding on the coat hanger as it slowly turned? What if your whole *world* was riding on the coat hanger? What would *you* see happening to the direction of the swinging weight?

Okay, then, how can this swinging weight prove that the Earth does, in fact, rotate on its axis?

Rock 'N' Roll Double Puzzle

What was the first rock group to appear on Earth after the disappearance of the dinosaurs?

If you can find the right word or phrase to fit each of the eight spaces below, you'll be halfway to your answer.

1. The Earth's crust is also called the

[_] __ __ [_] __ __ __ __ __ __ [_]

2. Made up mostly of molten basalt, this is sandwiched between core and crust.

__ __ [_] __ __

3. The movement of the Earth's crust is called

__ [_] __ [_] __

[_] __ __ __ __ __ __ __ __

4. Heat is dissipated from the Earth as plates are pushed down and "melted" in a process called

[_] __ __ __ __ __ __ __ __ __ [_]

5. Winds blowing from west to east at about 450 miles per hour make up the

__ __ __ [_] __ [_] __ __ __

6. Earth began accretion about 4.6 of these years ago.

__ __ __ __ __ [_] __

7. Earth's magnetosphere protects us from harmful

__ __ __ [_] __ __ __ __ [_]

8. The movement of weather systems based on the rotation of the Earth is called the

__ __ __ __ [_] __ __ __

[_] __ __ __ __ __

Okay, do you have all that figured out? Anxious to find the answer to the burning question? (No? Where's your sense of adventure?) All you need to do now is take all the letters that ended up in the brackets [x], and unscramble them.

What was the first rock group to appear on Earth after the disappearance of the dinosaurs?

GENERAL MAINTENANCE

Please note: Your OMICRON IV should operate flawlessly for several billion years. Self-correcting systems have been installed that should handle most serviceable situations. Nevertheless, certain areas of operations might require periodic maintenance. With basic care and common sense, the problems indicated below should never arise; if they do, however, maintenance procedures should be initiated without delay.

I.

Accumulation Of Greenhouse Gases

There are many ways you could look at the difference between the "twin sisters" Earth and Venus. But no matter how you cut it, the difference between this garden-planet and that orbiting disaster is going to come down to one thing: heat.

True, Venus is hotter because it's closer to the sun. That, however, doesn't begin to account for temperatures hot enough on the surface to melt lead.

Venus is a victim of both its atmosphere and the phenomenon called the greenhouse effect. If Venus had no atmosphere, its surface temperature would represent a balance between sunlight absorbed and internal heat radiated into space.

But an atmosphere changes everything. With an atmosphere, some sunlight is blocked out. On the other hand, internal heat is not as easily radiated into space. So, what is it in Venus' atmosphere that causes such terrible temperatures? Carbon dioxide. CO_2. The same carbon dioxide that we are releasing into our atmosphere in unprecedented quantities.

Carbon dioxide traps heat and reflects it back to the planet. Earth always has had less carbon dioxide in its atmosphere than Venus, because much of our carbon, present in great quantities as the planet was forming, was "trapped" in carbon-based rock. Plant life also helped keep CO_2 in check through the process of photosynthesis.

In recent centuries, though, we have mined the carbon from our planet and burned it as fuel, releasing more and more CO_2 into the atmosphere. At the same time, deforestation gets in the way of the limiting effect of photosynthesis. Bottom line? CO_2 is building up in our atmosphere.

And our planet is getting warmer.

If temperatures continue to climb as they have been in recent decades, the effects for those of us who are passengers on Spaceship Earth will be severe. For example, two principal areas of food production in the world (the American Midwest and the Ukraine) are likely to become

> **"Number one, the Earth is presently warmer than at any time in the history of instrumental measurements. Number two, the greenhouse effect is probably the principal cause of the current global warmth."**
>
> *James E. Hanson, Director of NASA's Goddard Institute for Space Studies, in testimony before the U.S. Congress*

desert land. Not only that, but the melting of glaciers and polar ice will raise the sea level and flood the world's coastal cities.

Carbon dioxide is not the only greenhouse culprit. Other chemicals such as chlorofluorocarbons (CFCs) also are greenhouse gases. Even water vapor is a greenhouse gas. As our planet heats up, more water vapor is absorbed by the atmosphere, further heating the planet.

Our use of coal, oil, and natural gas releases CO_2. The annual burning of African grasslands releases CO_2. The continued burning of the Amazon rain forest releases CO_2.

Controlling and reversing the greenhouse effect will require international cooperation.

II.
Depletion Of The Ozone Layer

Ozone is a gas that makes up only a small part of Earth's atmosphere. Small, but important. Ozone reflects most of the ultraviolet rays Earth receives from the sun.

Ultraviolet light is what gives you a tan on the beach. It is also what gives you a sunburn if you aren't careful. Too much ultraviolet can cause eye problems, skin cancer, and damage to crops and animal life — even life in the oceans.

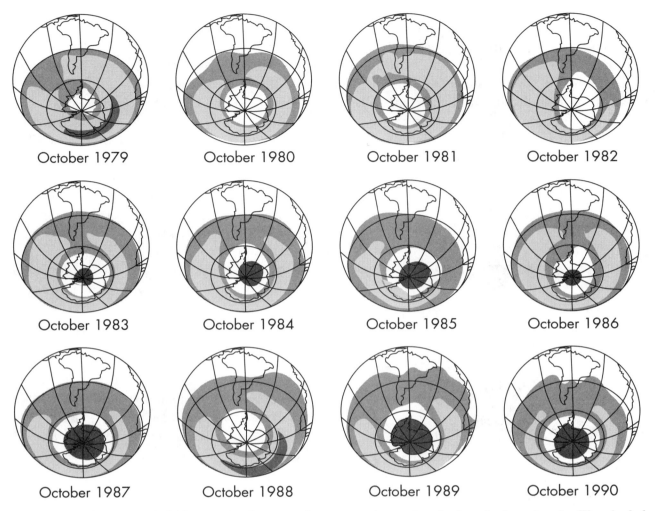

October 1979   October 1980   October 1981   October 1982

October 1983   October 1984   October 1985   October 1986

October 1987   October 1988   October 1989   October 1990

Satellite studies have revealed changes in the ozone layer over Antarctica during the last decade. The shaded areas show thinning ozone levels.

Recent research, brought about by satellite photography of the Earth, has revealed a general thinning of the ozone layer around Spaceship Earth. The thinning has become most severe over the Antarctic region, giving rise to what scientists call an "ozone hole."

Ozone depletion is caused by several things, and some fluctuations in the thickness of the ozone layer are natural. However, CFC gases absolutely devour ozone. CFCs, when they rise into the upper atmosphere, are broken apart by ultraviolet radiation. CFCs become chlorine, fluorine, and carbon. The carbon (as you know) is a greenhouse gas — bad enough. But it's the chlorine that destroys ozone.

CFCs are used as coolants and propellants. (Translation: They are used in the cooling coils of refrigerators and air conditioners and were used as the gas that makes spray cans spray — although other, safer propellants are now used instead.) CFCs also are used to make insulation material and foam packing.

To be honest, an ozone hole over the Antarctic region might not be such a big deal. Peculiar weather conditions and cold temperatures may thin ozone even without human "help." However, studies of ozone layers over the Antarctic have shown that these already low levels are still decreasing — and have been throughout the 1980s and 1990s. Other satellite studies have indicated ozone thinning over areas of Europe and the Pacific Ocean as well.

Millions of tons of CFCs already have been released into the atmosphere. However, scientists are working to find other materials that will do the things CFCs do, and many countries (including the United States) have banned CFC propellants. Protecting the ozone layer, like controlling the greenhouse effect, is an important global issue.

ACTIVITIES

DESIGNING A SOLAR COLLECTOR

YOU WILL NEED:

2 low, flat boxes (the boxes most clothing comes in are terrific, but shoe boxes will work in a pinch)

aluminum foil (the wider, heavy-duty kind)

black paint

a cup or bowl

a thermometer

There are several ways to take energy from the sun and convert it to electricity. Solar panels convert solar radiation directly to electricity, and you can purchase small solar panels at science and hobby stores, along with directions for their use. Heck, your pocket calculator probably already has one. Experiment to see if you can use these small panels (some may be sold as small wafers a few inches in diameter) to power instruments that usually run on batteries. (No, don't take your calculator apart. Not unless you are giving up math forever — along with balancing your checkbook.)

Not about to spend hard cash on a solar wafer? No problem. Here's one solar collector you can create on your own. (By itself, saving you the price of this book.)

Find a sunny spot in the back yard (or on the roof, if you can do that without the neighbor dialing 911). Line one of the boxes with a single sheet of aluminum foil (shiny side up). Make sure you fold the corners so that water won't leak out. Take another sheet of aluminum foil and paint it black. Line the other box with this sheet, black side up.

Set both boxes in the sun. Pour equal amounts of water into each box (1 pint should do it). Set aside another pint in a cup (also in the sun).

Wait an hour. (Re-read a back issue of *ODYSSEY* or watch one of the episodes of *Beakman's World* you have on tape.)

Measure the temperature of the water in the cup and record it. Pour that out and pour the water from the shiny aluminum foil box into the cup and measure its temperature. Discard that and measure the temperature from the black box.

Check your results.

Okay, now, what does all of that tell you about collecting solar energy? Which method captured the most heat? Why?

Next step, what could you do with what you've learned? It's nice to make warm water, maybe for a cup of cocoa — assuming you don't mind drinking water that's been sitting in a box with black paint. But...how could this process (or one like it) be used to make electricity? How can you convert heat energy into electrical energy?

LIFETIME ACTIVITY

What is one thing you could do to help keep Spaceship Earth in good working order? Walk along a street and pick up trash, plastic, aluminum, etc. for recycling? Terrific! That not only helps save resources, it also makes the street look nicer.

Okay, then, what's another way?

And another?

You may have seen such books as *50 Ways to Save the Earth* — there are several of them "out there." Make your own list. If you run out of ideas, ask someone else for their ideas to add to your list.

And then do some of them.

And ask two friends to do some.

And ask them to ask two friends to do some.

Spaceship Earth is our home. It's the only home we have. Every once in a while, it needs cleaning.

PARTS AND SERVICE

Assembly instructions for several of OMICRON IV's auxiliary parts are included with the basic planetary model. These parts are all easily assembled using those tools available to any sentient landlord species with opposable thumbs. If any questions or problems arise regarding assembly, a list of 20th-century service representatives is included.

I.
*T*elescopes *A*nd *O*bservatories

People have been building, using, and improving telescopes for nearly 400 years; but even the high-tech versions of today are based on two designs proposed in the 1600s. (In 1609, Galileo constructed his first refracting telescope, based on a design used in the Netherlands since the 1580s. In 1668, Newton invented the reflecting telescope— but hey, you already know this stuff.)

Telescopes have to do three things in order to work effectively. First, the telescope must gather light. The more light gathered by the telescope, the brighter a distant (and faint) object will appear. Second, the telescope has to magnify. Lenses are used to bend the light reflected off an object so that it appears larger. Third, the telescope has to make details visible that are indistinguishable to the naked eye.

Both refracting and reflecting telescopes have problems, however, that are "built in" to their basic designs.

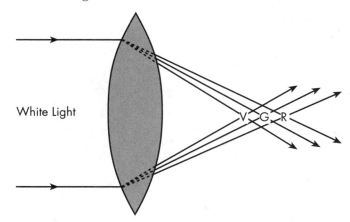

White Light

V G R

Check out that refractor design above for a minute. See the convex lens? (Convex lenses bulge out at the center. Lenses that are narrow-

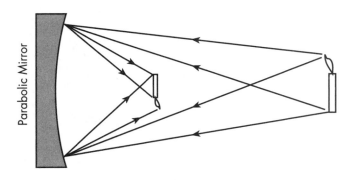

Parabolic Mirror

One way to minimize chromatic aberration is to increase the focal length of the lens. This, however, limits the size lens that can be used.

er at the center are called "concave.") The problem with a refracting telescope involves that convex lens. Lenses not only bend light, they also act as a prism, separating light into the colors of the spectrum. Violet rays are bent more sharply than red rays. The end result is that, when you look at a distant object through a refractor, there is a colored red or violet haze around it. Astronomers call this problem "chromatic aberration," although in private they probably have stronger words for it.

ARE YOU DYING TO SPEND SOME TIME IN A MAJOR OBSERVATORY?

The Lick Observatory in California is home to a refracting telescope. The telescope is mounted on a pillar. And the pillar?

That contains the remains of James Lick, who paid for the construction of the observatory and then "bought it" himself in 1876.

Modern astronomers also have minimized chromatic aberration by use of a two-piece lens made of two different materials. Dense glass (flint glass) is sandwiched to the convex lens and works much like a pair of glasses to refocus the separated colors onto one point.

A lens refracts. A mirror reflects. Reflected light doesn't break apart into colors. If you don't use a lens, you don't have chromatic aberration.

Bingo! Isaac Newton designed his telescope based on a concave mirror, used much like the lens, to bring light together to form an image. The light reflects off a concave (or parabolic) mirror and onto a second, small mirror. The small mirror reflects the light into the eyepiece.

Sound perfect? When Newton designed this telescope, nobody was able to build it. The technology did not exist to make a mirror without

slight flaws in the surface — and in a reflector, every flaw is magnified (just like the stars). It wasn't long, however, before the making of mirrors became more exact, and reflecting telescopes became the "scopes of choice." Most large observatories use reflecting telescopes. Even the Hubble Space Telescope is a reflector.

So why do people still use refractors? There are advantages. The small secondary mirror in a reflector has to be positioned in the middle of the tube, getting in the way of light entering the scope. For wide-field viewing (of star groups, for example) this is not a problem; but for single-object viewing (of planets, for example) this can cause some fuzziness in the image. Most observatory viewing is on deep-space objects, so reflectors are preferred; but there's definitely still a place for a good refractor.

YOU WILL NEED:

- the cardboard insert from a roll of paper towels (or a wooden dowel at least 1/2-inch thick and 1- to 2-feet long, for the more daring builder)
- a small square board, about 1-foot square
- a protractor
- a thin marking pen
- a plastic straw
- an empty spool of thread
- a nail (and a drill if you are that same daring builder)
- a string with a weight tied on the end
- a thumbtack

A C T I V I T Y
BUILDING A THEODOLITE — UH-HUH

You want to buy a telescope? That can be complicated (and expensive). Why don't you just cool your jets for a minute.

An ancient (and reliable) way to locate a star or event in the night sky is by measuring its ALTITUDE and its AZIMUTH. The altitude is the number of degrees above the horizon, and the azimuth is the number of degrees to the east or to the west of north.

Find the approximate center of the flat board and mark it. Set the protractor so that the center of the straight edge is on the center mark. Draw around the semicircle and mark every 10 degrees up to 180 degrees. Flip the protractor over and complete the circle, marking degrees from 190 degrees to 360 degrees (which is also zero degrees). Label the zero degree mark "NORTH."

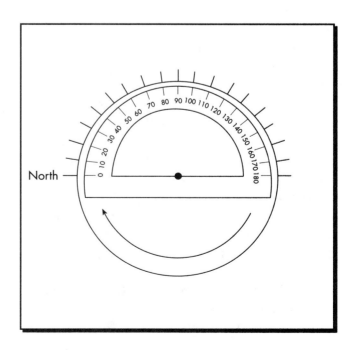

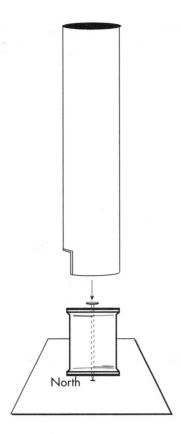

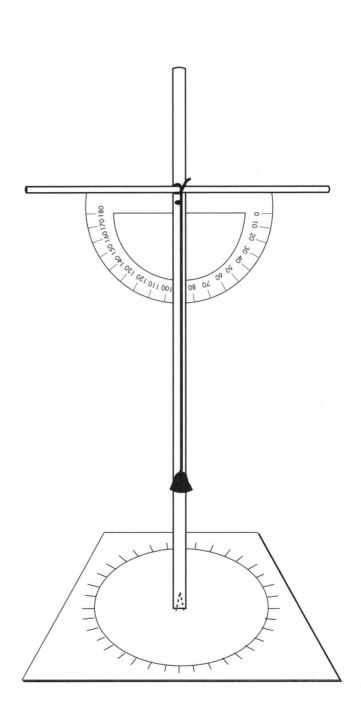

Insert the nail through the center of the spool and pound it into the board at the center mark.

Mark a notch at one end of the paper towel tube and set the tube over the spool. Line up the notch with north.

Daring Builder Version

Instead of the above, mark the center, then hammer the nail so that it just begins to come out the other side. Flip the board over and, using the nail's point as center, mark off the same 360-degree circle mentioned above. Now pound the nail all the way through so that it emerges at least an inch on the other side.

Drill a hole down the center of the dowel. (You will need a vise or some help for this, and you should be prepared to cut an inch or two off the dowel and try again if you mess up. Hey, you wanted to be daring!) Now set the hollowed out dowel over the protruding nail. Mark a spot on the base of the dowel and line it up with north, as above.

For Everybody, Daring or Not

Tie the string around the center of the drinking straw and tape the straw to the straight edge of the protractor. The weight on the end of the string should hang down about 6 or 7 inches and move freely.

Push the thumbtack carefully through the plastic protractor, in the middle of the straight edge section. Finally, attach the protractor to the upright support by pushing the thumbtack into either the paper towel roll or the wooden dowel.

Voila!

So now you're thinking, "What is this weird-looking thing I just made that still has a few drops of my hard-earned blood on it!"

Simple. It's a theodolite. Uh-huh.

How does it work?

Take it outside on a starry night and set it on a flat surface. (How about the picnic table your dad keeps saying he's going to lug into the garage for the winter but never does?) Turn the base so that "North" points north (find Polaris — compasses are for wimps).

Now, find a star you would like to chart. Turn the paper towel roll so that the end of the straw that is taped to the 180-degree end of the protractor points in the general direction of the star. Bend down and carefully move the straw so that you can see the star through it from the other end (the zero degree end).

When you have sighted the star, hold the straw steady and look at the string. What degree mark on the protractor is the string hanging over? Subtract the number from 90 and write down your answer. That will be the star's ALTITUDE. (Helpful hint: If the string is hanging over a number that is greater than 90 degrees, you are looking through the wrong end of the straw. Be very cool and nod knowingly as you turn the whole contraption around and start again.) Now look at the mark at the base of the towel roll. Mark the degrees indicated there. That is the star's AZIMUTH.

Using altitude and azimuth, you can locate any object in the sky. In your record-keeping, also note what time in the evening you saw this star, since Earth's rotation will keep that baby on the move.

Buying a Telescope

Did the above activities make you forget about buying a telescope? No?

All right, then, here are some tips.

1. Telescopes aren't cheap. For a good "starter" scope, figure on spending a couple hundred dollars.

2. Decide what type of scope you would like. Refractors are better for viewing planets and single objects. Reflectors are better for viewing star clusters, nebulae, and other "wide" area objects. (But don't worry, either scope will allow you to see all of these things.)

3. Have you considered a good pair of binoculars? Good binoculars allow you to see nearly as much as a starter scope, and might be more comfortable to use (after all, your brain is more used to seeing through two eyes). If you want binoculars, you should get a pair with a front lens diameter of about 50 millimeters and

a magnification power of 7x. Look for a label that reads "7x50." They will probably cost about $200. One last thing about binoculars. They are easy to carry around and are useful for things other than astronomy — so think about it.

4. Don't be swayed by the promise of 1,000x magnification (or more). At these high magnifications, the images you see are often faint and fuzzy. A good scope should allow you magnification from 50 to 400 times (tops) — more than that you will seldom use.

5. If you want to take pictures, make sure you have a scope that will accommodate a camera.

6. A tripod is nearly as important as the scope. Make sure it is portable, easy to use, and steady.

7. And, finally, before you buy anything, you are going to want to ask a lot of questions. That means you have to buy your scope from someplace other than the local department store. You have questions, you want answers. A $200 to $300 telescope might not be the biggest purchase you've ever made, but it's big enough. You don't want to make that purchase from the guy who worked in the shoe department last week.

II.

*T*elescope *L*ocations

So, where do you find big-time telescopes, anyway?

In observatories.

There are several large observatories for optical telescopes. Famous international observatories include the following.

The Observatory of Paris, France

The Royal Greenwich Observatory in England (official location of the Prime Meridian —

zero degrees longitude — just in case you were wondering)

Cerro Tololo Observatory in Chile

Russian Astrophysical Observatory in the Caucasus Mountains

British William Herschel Observatory in the Canary Islands

European Southern Observatory in Chile

British Infrared Observatory in Mauna Kea, Hawaii

Canada/France/Hawaii Observatory in Mauna Kea, Hawaii

Zelenchukskaya Observatory in Russia

Siding Spring Observatory in Australia

Pulkovo Observatory in Russia

La Palma Observatory in the Canary Islands

The German Academy of Sciences Observatory in Berlin

The Dominion Observatory of Canada in Ottawa, Ontario

The Dominion Astrophysical Observatory in Victoria, British Columbia

The David Dunlap Observatory in Richmond Hill, Ontario

Of course, there are several famous optical observatories in the United States as well:

Kitt Peak Observatory in Arizona

Keck Telescope Muana Kea Observatory in Hawaii

Mt. Wilson and Palomar Observatories in California

Agassiz Station Observatory in Massachusetts

U.S. Naval Observatory in Arizona

Lick Observatory in California

Yerkes Observatory in Wisconsin

McDonald Observatory in Texas

U.S.-owned Cerro Tololo Observatory in Chile (that's cheating a bit)[*]

---

[*] "Observatories," *Encyclopedia Americana*. Danbury, CT: Grolier, Inc.,1992. Vol. 20, pages 600–605.

WHAT'S THE MOST EXCITING THING HAPPENING IN ASTRONOMY TODAY?

"Without a doubt, the Hubble Telescope."

Astronomer Vera Rubin

That's a lot of sky watching. But the most famous optical telescope can't be found in any of those observatories. The current cutting edge is "up there" — the Hubble Space Telescope. (Actually, the Hubble is only 576 kilometers "up there." The moon is 1,000 times farther away than that.)

Launched into orbit from the space shuttle *Discovery* on April 25, 1990, the Hubble Space Telescope (HST, among friends) looked at first like it might be a billion-dollar loser. The mirror was flawed, the solar panels wouldn't deploy properly, and antenna systems weren't working. The HST didn't get off to a great start.

Nevertheless, weak and nearsighted, HST was still able to see more and see farther than any Earth-based scope. The reason was simple. In space there is no dust, no water vapor, no pollution, no street lights, no jet planes — no atmosphere.

The icing came in 1993, when a NASA shuttle mission caught up with HST and made repairs. During three space walks, HST was brought back into shape — it was even given a pair of glasses (okay, it was a corrective lens) to counter the flaw in the main mirror. ("Mirror," so it's what kind of scope? Right, a reflector.)

New images are coming in from HST faster than scientists are able to analyze them, but it didn't take long for Hubble to find itself right in the middle of a cosmic mystery.

The Hubble Constant Cosmic Mind-Boggler

One thing the Hubble Space Telescope is supposed to be good at is looking to the most distant (and therefore oldest) places in the universe. Distance is a tough thing to measure in space. If you want to know how far away Barnard's Star is, for example, you can't exactly hop in the old sports car and drive. (At a steady clip of 100 kilometers per hour, that trip would take you 64,000,000 years one way — assuming you don't stop for breakfast.)

Yet, if we can tell how far the most distant galaxies are from us and how fast they are moving away from us, we can compute backward and tell how old the universe is. Pretty cool, huh?

There's even a formula for all of this (you knew there would be): H = V/D.

The speed at which an object is moving away from us (velocity) divided by its distance from us will always give the same number. In this case, that number is "H," also called the Hubble Constant.

Distance, in this case, also means age — since a galaxy that is 4 billion light-years away must be 4 billion years old (it took 4 billion years for the light from that galaxy to reach us). For this formula, however, distance is measured in megaparsecs (3.26 million light-years = 1 megaparsec).

The HST has been able to measure distances and velocities from enough locations to have a pretty good idea of what the Hubble Constant should be — and it should be somewhere around 80.

And since we know nothing can travel faster than the speed of light (300,000 kilometers per second), we can get a good idea of how old the universe can be — max.

Can't we? Think about it while you warm up your calculator.

H=V/D

Distance is in megaparsecs

3.26 million light-years = 1 megaparsec

Nothing can go faster than light

Light moves at 300,000 kilometers per second

So now, what is the oldest the universe can be?

Bonus: Astronomers have identified globular star clusters estimated to be 16 billion light-years away. Any problems with that?

Stretch that brain.

One-Eyed Man-in-The Moon

Sketch of Mare Orientale

III.
*C*ameras *A*nd *I*maging

Seeing is believing. But even things we can't see are believable. Cameras and computer imaging give us the chance to see things we don't usually see or can't see, which means they are now believable, too. See?

Cameras take us where our eyes can't go, and that is especially true in astronomy. We only see one side of the moon, but cameras allow us to see the rest. Otherwise, we wouldn't be able to see Mare Orientale.

Photographs can be enlarged and enhanced to show details we could never see on our own.

Cameras can show us the surface of Mars or the rings of Uranus. And photographs can be kept, studied, and compared over time.

Compared to what? Just ask Clyde Tombaugh. In 1939 he was examining photographic glass plates of the night sky using a machine called a blink comparator. The comparator allows you to see plates of the same section of sky made on different nights.

"I saw a little image popping in and out," Tombaugh said. "Then I looked to one side and saw another image doing the same thing on the other plate."* There were several checks to be made, but Tombaugh had done something only two other people had ever done (and the list is not likely to grow). He had discovered a planet — Pluto. (Uranus and Neptune are the only other planets to have been "discovered," since the others have been recognized since before recorded history.)

WELL, HE EARNED IT

Between 1929 and 1945, Clyde Tombaugh spent 7,000 hours looking at plates through the blink comparator.

Tombaugh is the only skywatcher to discover a planet with a camera.

Cameras are also more versatile than our eyes. We can see white light or "whole light," made up of all the colors of the spectrum.

But that phrase is kind of a "white light" lie.

There are many more colors of the spectrum that we can't see. (They aren't referred to as "colors" because sooner or later some doofus would ask what they looked like.) Actually, they are wavelengths. Each color of the spectrum has a wavelength. When energy moves at that wavelength, we see that color. Red wavelengths are longer than blue wavelengths, and that is really

* "Galileo Encounters Earth & Venus." *The Planetary Report*, March/April 1991, page 15.

what makes the difference in those two colors. While our eyes are able to see only the wavelengths of the visible spectrum (red, orange, yellow, green, blue, indigo, violet), other wavelengths exist.

ACTIVITY
COLORS OF THE SPECTRUM USING A PEN

You've seen experiments where light is broken down into the colors of the spectrum. And, of course, you're expecting to be told to set up the same experiment yourself by pulling out your handy prism — yeah, like everybody just happens to have prisms jangling around in his or her pockets.

Maybe you do.

Do you have one of those cheapo ball-point pens with a cartridge inside a clear, plastic shell? (Does the person sitting next to you have one? Can you distract him or her for a minute?) Pull out the ink cartridge and you now have a cheapo prism.

Find a sunny spot, set the pen cartridge in the sunlight on a piece of white typing paper. Do you see the swirls of colors in the shadow of the pen? There's your prism! (Okay, so it isn't perfect — but what did it cost you?)

What exists at wavelengths even longer than red? Infrared is the term for these long wavelengths, and even longer ones are radio waves. But infrared and radio waves are invisible to us. Wavelengths too short for us to see are called ultraviolet. Even shorter wavelengths are X-rays, then gamma rays, and so on.

There's a lot going on that we can't see. Some cameras photograph other wavelengths and reproduce them within our spectrum in a print we *can* see.

The HST (always keep track of initials) has more than a reflecting telescope on board. There is also a high-resolution spectrograph, which can photograph objects for certain, specific wavelengths — both visible and invisible.

Why? Funny you should ask.

Spectral analysis provides a "fingerprint" for stars. In 1802, William Wollaston discovered that different stars presented different patterns of bright lines and black lines when analyzed from a spectrograph. Annie Jump Cannon, one of Harvard's computers, established...

(— one of Harvard's WHAT?)

Don't interrupt! Cannon organized the spectroscopic analysis of many stars and created a stellar classification system that is still used. She then went back and began a systematic classification of stars — 250,000 in all — which were collected into a catalog used by astronomers today.

(— one of Harvard's WHAT?)

(— computers!)

From 1877 to 1919, the painstaking work of collecting and organizing information fell to a group of women. The idea was that women would be more patient than men in going over massive amounts of data. They plotted orbits of planets and asteroids, studied photographic plates, organized catalogs of stars, and examined their brightness. Harvard's "computers" not only paved the way for women in astronomy, they also compiled the information necessary for many of the discoveries of their time.

But spectroscopy does not just examine visible light, spectrographic cameras can also record invisible wavelengths.

For example, infrared studies reveal heat sources, and can locate clouds of dust in space that are heated by stars within. Infrared rays also penetrate clouds and so give us pictures of the surface of Venus even through the clouds.

Near infrared photos of Venus penetrate only some of the clouds, allowing us to "slice off" the upper layers and see the lower clouds that make up planetary weather patterns. Radar and X-ray studies allow satellites to find uranium beds on Earth, water beneath the Sahara Desert, and mountains under the oceans, and even monitor the condition of our rain forests and oceans.

Even radio waves (very long wavelengths) can reveal secrets.

Radio waves also travel through space, and radio telescopes can pick them up just as optical telescopes can pick up light. Thomas Edison (the light bulb guy) suggested that radio waves might come from the sun, as did English physicist Joseph Lodge. Several scientists later, radio waves had been recorded, not only from the sun, but also from the center of the Milky Way galaxy.

Radio telescopes have huge dishes that serve as antennae, collecting radio waves and bouncing them to a central receiver (not too different in design from the reflecting telescope). Radio telescopes are humongous because radio signals are very weak.

The Arecibo radio telescope in Puerto Rico is the world's largest. Its dish is 1,000 feet across and can pick up signals as weak as $1/100,000,000,000,000$ of a watt.

SOME THINGS ARE IMPORTANT BECAUSE YOU DON'T FIND THEM

No camera on any satellite using any photographic technique has ever been able to photograph a political border.

HOW WEAK?

If we collected all the energy received (in radio waves) from one quasar galaxy by one radio telescope for the entire life of that telescope, it would not be enough to light a 20 watt bulb.

A C T I V I T I E S

GRAPHING SPACE PHOTOGRAPHS

As you might have guessed, photographs from space aren't actually sent as photographs. Finding a drug store on Saturn to develop the film is a real pain. Instead, pictures are "taken apart," block by block, and signaled back to Earth where they are "reassembled."

```
┌─ ─ ─ ─ ─ ─ ─ ─ ─ ─ ─ ─ ┐
│ YOU WILL NEED:
      some graph paper
│     a piece of carbon paper
      a picture to send
└ ─ ─ ─ ─ ─ ─ ─ ─ ─ ─ ─ ─ ┘
```

Imagine a picture on graph paper. Each square is either light, dark, or gray. This becomes the signal "1" (light), "2" (dark), or "3" (gray). Computers use full color and are much more complex than this, but the idea is the same.

So, instead of sending a photograph, you could send a series of numbers that someone could reconstruct on another piece of graph paper.

Try it. Find a picture and copy it onto your graph paper using the carbon paper. (Don't get too big a picture, or you'll be stuck on this page for years and won't get to read the rest of the book until it's out of date.) Record a signal (1, 2, or 3) for each square on the top row. You will need a special "code" signal to indicate the end of one row and start of another. How about "0"? Record signal numbers for every row in your picture, and give them to someone. If that someone has a piece of graph paper like yours, he or she should be able to reconstruct your picture.

Find out.

Feeling daring? Try adding colors by adding code numbers.

(Aren't you glad you don't have to do this for a living? Aren't computers wonderful?)

CAPTURE INFRARED REDHANDED

So, you don't believe in infrared, eh?

YOU WILL NEED:

a piece of white typing paper

a thermometer you can ruin and not get punished for it

a prism

You are about to record some small (but significant) changes in temperature, and for that, you will need to paint the bottom of the thermometer black. (Why? Because the black will absorb more energy and amplify the difference in heat.)

Set up your typing paper and prism so that sunlight passes through the prism and is broken apart into a spectrum on the paper.

Set the thermometer on the white paper well outside the spectrum area and record the temperature after at least three minutes.

Move the thermometer to the blue end of the spectrum, wait, and record the temperature there.

Move the thermometer to the red end and do the same.

Now move the thermometer just beyond the edge of the red part of the spectrum. If there was no such thing as infrared, this last temperature would be the same as the first one — on

the white paper.

But it isn't. Like it or not, something *is* there — and it's called infrared.

FOR THE ULTIMATE IN ENERGY WEIRDNESS, TRY THIS:

When you think about it, it's kind of strange that our eyes see one band of energy wavelengths while our ears hear another band of much longer wavelengths. Apparently, the only difference between color and sound is the length of the energy wave.

Hmm…

What if we could *see* sound and could *hear* color? What would the cafeteria *look* like? What about a rock concert? What would a rainbow *sound* like?

Let your imagination work on that for a minute, and write a description of a world where these two senses are reversed — or maybe even where they cross over.

IV.

*R*ockets *A*nd *R*obots

Telescopes are terrific instruments for investigating space, but there's no substitute for being there. And you can't get there without rockets.

Rocket history goes way back to the ancient Chinese (remember the rocket-powered arrows?), but the development of rocket technology is really a 20th-century project.

As usual, the dreamers came before the builders. Konstantin Eduardovich Tsiolkovsky

aloft less than 5 seconds and didn't look like much, but it worked.

Other scientists, particularly in Germany and Russia, experimented with rocketry. Sadly, it was World War II and the quest for greater weapons that really made rocket science a priority. Germany made progress in rocket stability, accuracy, and distance. When German rocket specialist Wernher von Braun came to the United States after the war, the U.S. rocket program went forward on full thrusters.

published *Dream of the Earth and Sky* in Moscow in 1895 and followed it with *Exploration of the Universe Space by Reactive Apparatus* in 1903. He may not have been much for titles, but Tsiolkovsky explained the principles by which rockets could (and later would) take us from the confines of Spaceship Earth. He also recommended use of liquid propellants, overcoming one of the early stumbling-blocks of rocket experimenters hung up on solid fuels. He also designed multi-stage rockets and proposed using a gyroscope for stabilization in space. (He also described how space stations could be used for minimum-gravity assembly of equipment.)

America's early pioneer in rocketry was Dr. Robert H. Goddard. Goddard was one of the first to take Tsiolkovsky's ideas and make them work. After working with solid-fuel rockets, Goddard launched the first liquid-fuel rocket in March 1926. It only flew 56 meters and stayed

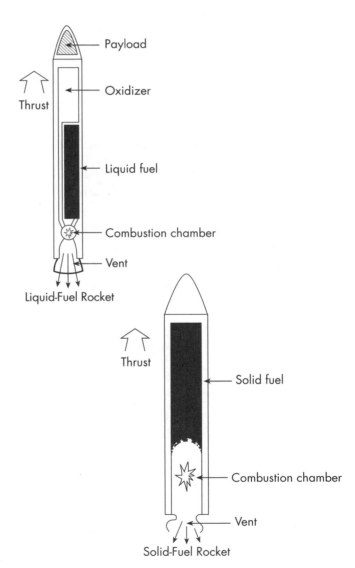

ACTIVITY

BUILDING A CHEAP ROCKET

Rockets move in one direction because they are responding to an equal force in the other direction. It's not the exhaust pushing against the launch pad that pushes a rocket into space, it's that every action (like the exiting force of the exhaust) has an equal and opposite reaction (like the rocket lifting skyward). In other words, try this:

YOU WILL NEED:

two balloons
a soda straw
some thread
some tape
a paper cup

Cut the bottom out of the paper cup. Blow up one balloon (if one of your balloons is a long one, use that one here), and use the narrower end of the paper cup as a stopper. Do this by folding over the end of the balloon and tucking it into the paper cup, which should hold the folded end in place and keep the air from leaking out.

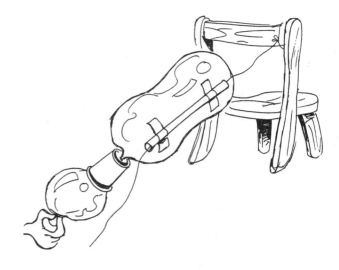

Insert the thread through the straw and either tie one end to something or have a friend hold it. Tape the straw to the inflated balloon in two places, with the end of the balloon facing you. (Be careful to keep the cup in place.)

Inflate the second balloon and stuff it carefully into the cup with the open end facing out. Inflate a little more. Ready? Now, tilt the thread slightly down from where it is tied and let go of the end of the balloon.

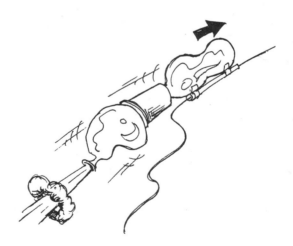

Why does it move UP the thread? Because every action (the air rushing from the balloon) has an equal and opposite reaction (the balloon climbing the thread).

It's a rocket. As a matter of fact, if everything went as planned (which may take some practice) it's a TWO-STAGED rocket. As the first balloon deflates and falls off, so will the cup, and the second balloon can take over. Well, that's the theory, anyway.

V.

*O*ur *I*nterstellar *T*ravelers

The two Voyager missions are among the most spectacular long-term events in astronomical history. Who knows when scientists will stop finding new things from the pictures we received from our solar system? But now that *Voyager 1* and *Voyager 2* have left our immediate neighborhood, most people think the mission is over.

Think again!

Both Voyagers are still out there, cruising along as ever. Communication systems are still functioning, and with luck, one of the craft will still be communicating in the year 2012 when it leaves the magnetic sphere of our sun, the heliopause. And while power and communications may fail soon after, the crafts will keep on going. Following current trajectory, here's a timeline with a hint of what is still to come for *Voyager 2* (V/2):

Year 8521 — V/2 passes within 4.03 light-years of Barnard's Star, a star that may have a planetary system of its own.

Year 21,319 — V/2 passes within 3.21 light-years of our closest neighbor, Proxima Centauri.

Year 26,262 — V/2 enters the Oort Cloud.

Year 28,635 — V/2 leaves the Oort Cloud. We can finally say it has left our solar system (assuming "we" are still here).

* Finnerty, Dan. "The Voyager Interstellar Mission." *The Planetary Report*, July/August 1989, pages 23-25.

Year 129,084 — V/2 will be closer to another star, Ross 154 (5.75 light-years), than it is to the sun (6.39 light-years).*

The development of rockets and satellite missions has gone hand-in-manipulative device with the development of robots. If some*one* can't be there, some*thing* can.

As a matter of fact, most satellites do so many different things on command you could get away with calling them robots. If robots in space are housed in satellites, then robots on the ground (or surface of some planet) would be housed in ROVERS.

We saw Dave Scott of *Apollo 15* driving the lunar rover, leaving patches and doing wheelies on the moon. Looked like fun. The Mars rover, however, will not require an astronaut at the wheel.

The Mars rover will need to be able to avoid or step over boulders, chasms, and whatever else Martian soil may offer. It will need to withstand extremes of temperature, which is one reason

researchers have tested parts of the rover in "slightly dormant" volcanoes and in Antarctica.

And it will need to think on its own, since there is a definite communications delay between Earth and Mars. The time it takes for a signal to reach Mars from Earth depends on how close the planets are to each other. At their closest, the "travel time" for a signal is about 5 minutes. At their farthest, it's more like 40 minutes. That means if the rover's cameras pick up something dangerous in its path, that picture must be communicated to Earth, where a decision must be made, a command given, and that command transmitted to the rover. At best, that means 5 minutes to receive the picture, a few seconds to program a response, and then 5 minutes more for the response to get to the rover. Let's see, that's, um, more than 10 minutes. Want to see how big a problem that is? Close your eyes and walk for 10 minutes. (No! Only kidding!)

Clementine wasn't a rover, and as a satellite, it was pretty simple. It might never be as famous as Voyager — but then again, it didn't cost much, either. *Clementine*, launched in January 1994, was designed to map the moon, search for water ice, and head off to explore an asteroid (Geographos, if you're curious) — and do it all on the cheap. The moon mapping went well, and *Clementine* found some evidence that there may be ice on the moon, after all. However, a computer glitch caused several engines to fire, spinning *Clementine* out of control. No asteroid this time.

And *Clementine* brought us back to space. The Voyager missions took so long, we convinced ourselves that we were still deeply involved in a space program. On the contrary, NASA's planetary exploration program fizzled out after the Voyagers were launched in 1977 and the Pioneer Venus missions in 1978. The next planetary missions weren't until the Galileo and Magellan projects in 1989.

It's been a long time since October 4, 1957, when *Sputnik I* beeped at us. Many of us have left the confines of our gravitational pull. One (astronaut John Young) has actually left six times. And 12 people have walked on the moon.

One thing is for certain. In the future, more of us will have the chance to step away from our Spaceship Earth and, for a while, live and work in space.

VI.
*S*ervice *R*epresentatives

This is not a complete list. Any of your listed service representatives should be able to offer you a complete listing, should the need arise.

Astronomers and Observers:
Clyde Tombaugh, Percival Lowell, Annie Jump Cannon, Henrietta S. Leavitt, Anthony Hewish, Williamina Fleming, Antonia Maury, Vera Rubin, Eugene and Carolyn Shoemaker, Fran Cordova, Mercedes Richards, Edwin Hubble, Jan Oort, Jocelyn Bell Burnell, Sidney Wolff, Carl Sagan

Cameras and Imaging:
Albert Einstein, Thomas Edison, Wilhelm Roentgen, Bradford Smith, Joseph Lodge, Karl Jansky, Zoltan Lajos Bay, Cecilia Payne-Gaposchkin

Rocketry and Robots:
Bruce Murray, Donna Pivirotto, Alexander Kemurdjian, Konstantin Tsiolkovsky, Robert Goddard, Max Valier, Johannes Winkler, Sergei Korolev, Wernher Von Braun, James Van Allen

People in Space:
Valentina Tereshkova (first woman); Sally Ride (first woman—American edition); John Young (most NASA missions); Guion Bluford (first African-American); Michael Jackson (first "moonwalker" — only kidding!); Neil Armstrong (actual first moon walker); Christa

McAuliffe (first teacher); William Thornton (oldest American, 56); Sally Ride (youngest American, 32); Yuri Gagarin (first in space); Alan Shepard (first in space — American edition); Beakman (farthest out)

A C T I V I T Y

MAKE YOUR OWN MESSAGE FROM EARTH

As you probably know, *Voyager 1* and *2* carry with them a recording from Earth. For whom? Who knows? Nevertheless, the recordings are out there, waiting to be found.

What if the content of those recordings was completely up to you? What would you say? Would you include any music? Let's say you had space for 200 words and two musical pieces. Write your message and be ready to explain your musical choices.

The Double-Trouble Mind-Boggler

Each question can be answered with two rhyming words that will shake up your memory, making this a _____ _____. (That's right, a "joggle boggle." Get it?)

1. What is an unshaven space telescope?
2. Where do you dock a comet?
3. What is a Martian dog cart?
4. What was Goddard's miniature launcher?
5. How do you shave a star?
6. Who sings at the moon?
7. What kind of verbal award did Tombaugh receive?
8. What part of a contract refers to the limits of the sun's magnetic sphere?
9. Who really "gets into" invisible light?
10. How do you transmit disease to light waves?

UN-BOGGLE THIS! NON-EARTHLING

If you were a whatsis from "out there," somewhere sometime in the future, and you happened to intercept *Pioneer 10* or *Pioneer 11*, you would know instantly that it came from some other intelligent species.

But what could you figure out other than that? When it was launched in 1972 *Pioneer 10* was equipped with a plaque – a picture from the mother ship (Spaceship Earth) that looked like the one on the top of this page.

ACCESSORIES

Your OMICRON IV is designed to remain functional despite the technological development of your species. We are proud to provide you with the planetary spaceship that grows with you. There should be no concern over obsolescence or the need to return to the manufacturer for upgrading. With OMICRON IV, the future is yours.

I.
*F*uturism

Futurism. Future science. Our eyes were put in the front of our heads so that we could look forward, and great thinkers have been doing that for thousands of years. Leonardo da Vinci sketched plans for a helicopter hundreds of years before such a device could be constructed. Futurist Alvin Toffler predicted the breakup of such large centralized governments as the Soviet Union 20 years before it happened. *Star Trek* was using sound lasers to heal 10 years before medical researchers began to experiment with them.

A FUTURIST is someone who studies the way things are and thinks of how they might be in the future. Futurists think this way by EXTRAPOLATING. Did you ever extrapolate? Sound disgusting?

ACTIVITY
CAN YOU EXTRAPOLATE?

Have your teacher collect the absentee lists from your school for the last month. Make a graph, with dates and days of the week along the bottom and number absent along the side. Count the number of students absent from your school for each day and plot it on the graph.

Now, think ahead. Make a mark on the graph *predicting* how many students will be absent each day next week. Think of *why* you are predicting the numbers you have chosen. There are probably *trends*. Which days of the week seem to have greater absentees? That is a trend.

As the week goes by, check your predictions. If you are close, pat yourself on the back. If you are way off, see if you can figure out what went wrong. And by the way, you are extrapolating.

> ## EXTRAPOLA-TION
> **To examine a current trend and extend it to a logical conclusion.**

II.
*L*iving *I*n *S*pace

Logically, we have a future in space — as a place to visit and a place to live. Why is that "logical"? Okay, here goes:

Our species has come to the point where we are *able* to enter space. If there is anything the 20th century has taught us about ourselves it's that if we *can* do something, we *will* do it.

Especially if it is in our own interests, and learning to live in space definitely is. We know, for example, that Earth's history includes several nasty collisions with asteroids or comets. Well-established colonies on other planets may one day be the key to our survival as a species. (Relax, this is long-range stuff — there are no comet collisions scheduled for next week.)

Also, our growth as a species requires space

and resources. The supply of both on Spaceship Earth is (like any spaceship) limited. But "out there"? Now *that's* elbow room!

III.

*T*he *S*pace *S*huttle *E*xperience

DO YOU GET PEANUTS IN FIRST CLASS?

The space shuttle can land on any runway at least 10,000 feet long (the size of the runways at most major city airports). However, most cities don't have the radio equipment necessary to guide the shuttle in, so such a landing would only be in an emergency.

Experience with the Apollo missions, *Skylab*, and the space shuttle have taught us that living in space involves overcoming serious biological problems. To escape gravity, a spacecraft must reach speeds of 32,000 kilometers per hour, Earth's escape velocity. At liftoff, acceleration and vibration cause disorientation and discomfort, and can affect critical reaction time. Once in space, near zero gravity causes problems with motion sickness, muscle tone, and nutrition. Not to mention the fact that weightlessness causes body fluids to rise, swelling your face like a blowfish.

This does not make for happy campers.

Low-"G" Is for "Gag."

Why is there motion sickness in space? Our sense of movement comes from subtle changes in fluid-filled canals in our inner ear. One canal lies in a vertical plane, another in a horizontal plane, and a third in a lateral plane. Fluid movement in all of them depends on gravity. Near-zero gravity? Near constant motion sickness. And believe this: The space shuttle is no place to hurl.

Most people who spend time in space are able either to get used to motion sickness or train themselves to avoid it. More dangerous for long-term voyages is the effect of microgravity on the human body. Astronauts have experienced nutritional loss, particularly of calcium; and while that's no big deal for a day or even a week, it can get pretty serious over a period of months or years. Programs of resistance exercise and special diets help minimize these problems.

You Can't Eat It If You Can't Catch It

Mealtime in microgravity is an adventure — but it isn't as bad as it was in the Apollo days. All your meals don't look like they come in toothpaste tubes (although some do). Meals come on a tray attached by Velcro to a table. A knife, fork, spoon, and a pair of scissors (to open containers) are held to the tray by magnets. Food needing rehydration is injected with water. (Can't pour water into your instant pudding and start stirring, that would be more like instant graffiti.) Food to be heated is set in a forced air convection oven. And the selection? Not too shabby. Here are samples of what you might find:

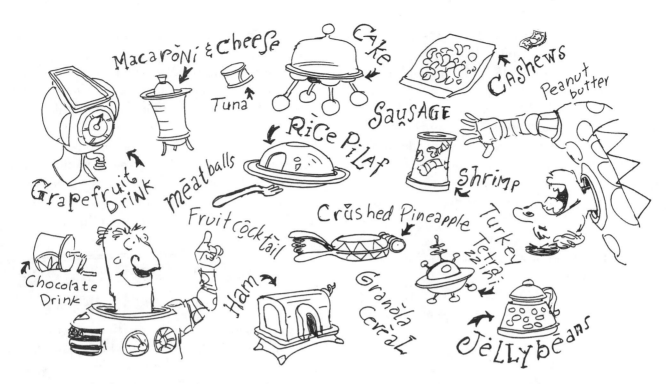

But, with all that eating, what do you do if, you know. How do you, well, like, everybody's gotta…

A C T I V I T Y

REDESIGN A SPACE SHUTTLE MENU

Is your favorite food missing from the NASA menu? It's time to fix that. Pick a favorite meal or dessert that's missing from the selection list and "redesign" it so that it can be included. Keep the following microgravity rules in mind:

1. Food must be dried at first. Hot or cold water can be added, but it must be injected, not poured.

2. Nothing that makes crumbs is allowed, unless the item is safely sealed. Nobody is going to be injured by a floating chocolate chip, but you wouldn't want it to float into the computer that controls life support, either.

3. "Sticky" foods are good because they will stick to spoons and forks, making them easier to eat.

4. Minimize waste. Can your food containers be used for anything after the meal?

So go for it. Design a meal and describe how it would be prepared and served.

But if you eat, well…you gotta…you know!

The Bathroom Plan: Toilet and Shower

Okay, okay…here's how you go to the bathroom on the shuttle.

The toilet (unisex — sorry, there are no men's and women's rooms here) is mounted on the wall (remember, sideways is okay in microgravity). As you sit, the specially designed seat collects both urine and feces. Rather than having this stuff float free, air is forced past the toilet seat, pushing any waste products to the bottom of a porous bag. This is sort of like a constant "air flush." When you're done, you need to remove the bag, treat it chemically, and store it. There is no dumping in space. (Remember that little fact when your mom or dad comes back from their first trip in space. Maybe you don't want to ask, "What did you bring me?")

And if that system fails, there are always good old-fashioned diapers. (Hey, you asked.)

Showering is hard work in space, and a nice refreshing shower is anything but refreshing. Completely sealed, the astronaut sprays himself or herself with soapy water, then must vacuum it off his or her body and off the inside of the shower stall. Most astronauts don't like the hard work setting up the shower and prefer a simple sponge bath — done slowly and carefully so that no water "escapes."

Z-z-z-z-z-z-z-z-z-z-z

Everybody needs to eat and use the toilet. But everybody needs to sleep, too — and sleeping is no piece of cake (sorry, no cake — too many crumbs).

These are the Z-Zone accommodations on the shuttle:

The shuttle holds four "rigid" sleep stations. Each contains a sleeping bag, space for storing personal belongings, a light, and a ventilation duct. Each bag is attached to a padded board. The sleeping bags are just tight enough to make the astronaut feel like he or she is lying against a mattress. Each astronaut also has a sleep kit, containing an eye cover and ear plugs. (After all, work on board the shuttle continues, even if you're zonked out.)

So does that make everyone cozy? Not exactly. There is still the problem of floating arms. Some astronauts keep their arms inside their bags, while others just let them float up toward their heads. And speaking of heads, a few astronauts have complained of "head nodding," apparently caused by the rhythm of the sleeper's pulse in the back of the neck. But, hey, if you're tired enough....

ONE SMALL BENEFIT

The position of your soft palate in your upper throat doesn't change as you move. In other words...in microgravity, you don't snore.

Everybody's a Weight Lifter

Exercise is very important in space. It's easy to get through the day without working your body physically, so aerobic exercise is even more important in space than it is on Earth. About 1 1/2 hours are set aside for each astronaut to work out on such things as the stationary bicycle (which can be adjusted without regard to gravity), the treadmill (with built-in resistance), and the pulley equipment (using springs as resistance tension).

Of course, these pieces of equipment are built with straps and harnesses to keep you in place. The space station *Freedom* is designed with a complete exercise room, with enough equipment to keep any astronaut from getting bored with the daily workout.

These experiments with living in space are only a part of the many research and satellite projects designed for shuttle crews. They shared their status as biological "guinea pigs" with frogs, eggs, flies, and (yes) guinea pigs. Much of the experimentation was in preparation for the next step to life in space.

IV.
*T*he *S*pace *S*tation *S*top-*O*ver

Working onboard the space shuttle tells us about the biological problems of space life, but shuttle trips are still just temporary trips. For long-term lessons, you need a space station.

The Soviet *Salyut* and the American *Skylab* space stations have provided the greatest opportunity for long-term survival in space so far. The new *Mir 2* Russian station and the proposed U.S. station *Freedom* will extend that opportunity into the next century — at which time the Statue of Liberty will *still* not need to be repainted.

Why not? So what?

V.
*S*pace *S*pin-*O*ffs

The Statue of Liberty was last repainted with a special paint that hardens into a ceramic finish and resists corrosion and abrasion. The paint was designed by NASA for use in the salty air of Cape Canaveral. No space program? The Statue of Liberty continues to corrode.

Ever 👉 see those ads for glasses that you can twist around like a pretzel and they still won't break? The metal came from NASA research for use on *Skylab*.

Temper Foam softens space seats — and dentist's chairs, car seats, and even athletic chest protectors. Freeze-dried foods were developed for astronauts. Blankets made of Mylar hold heat over sensitive parts in space and are also used by campers and rescue squads.

The space program brought many spin-offs to the Earth-bound passengers of Spaceship Earth — and many of those spin-offs came from work aboard space stations. Not convinced? Sample this list:

✔ self-contained breathing system for firefighters

✔ elastic dental braces

✔ advanced airplane engines

✔ cooling fans for computers

✔ laser surgery

✔ flame-resistant building and clothing materials

✔ shock-absorbent foam padding for athletic shoes and hiking boots

✔ automobiles that can be driven by "joystick"

✔ insulin infusion pump for diabetics

✔ stronger, lightweight wheelchairs and racing bicycles

✔ Magnetic Resonance Imaging (MRIs)

✔ voice-controlled equipment

✔ water-recycling technology and ideas

✔ scratch-resistant eyeglasses

✔ reading machines for the blind

When orbiting laboratories are established aboard *Freedom* and *Mir 2*, you can expect that list to multiply. Next time you look at your favorite pair of running shoes, think of shock-absorbent foam padding. Then think "space."

A C T I V I T Y

REPLACING GRAVITATIONAL FORCE

While some space spin-offs were a direct result of microgravity, a future generation of space stations will be built to avoid the problems of weightlessness. Finding some way to imitate gravity would solve a lot of problems for space travelers. And the most likely way to do that is to replace gravitational force with centripetal force. Here's how:

YOU WILL NEED:

a paper cup
some string (about 2 meters long)
a bathing suit

Punch two holes on opposite sides of the top of the paper cup, just below the rim where the paper is rolled over. Pull a piece of string through the two holes so that there is about 1 meter of string on each side. Fill the cup two-thirds full of water.

Now, as the cup sits there on the floor (or driveway, if you're smart), gravity is forcing the liquid to rest in the cup. In microgravity, that stuff would be floating all over the place already.

Grab both ends of the string in one hand and lift the cup with the other. Shorten your hold on the string until you can hold the cup and still keep the string taut. Ready? Okay, quickly start to swing the cup around in a circle over your head. One of the following is going to happen:

1. The string will break or the cup will rip, in which case you are glad you're wearing a bathing suit.

2. The cup will hit something, in which case you'll wish you had gone outside on the driveway.

3. A little water will spill, but the cup will start swinging and the water will stay in the cup.

Now, while you're figuring out how to stop this experiment without getting wet, consider what has happened. Centripetal force has replaced gravitational force. That means we can imitate gravity if we design space stations that rotate like that cup. (Oh, and there is no way to stop this experiment without getting wet, so just take it like a scientist.)

Engineers have figured that to imitate a comfortable gravity, you would need to rotate a bicycle-shaped space station 1.6 kilometers in diameter and 122 meters wide at a rate of 1 revolution per minute.

VI.
*C*olonization *I*s *F*or *K*eeps

But *can* we live off of Spaceship Earth for a long period — or even indefinitely?

Volunteers from Biosphere II in Arizona are trying to figure out the answer to that question. Biosphere II was designed to be a self-contained, completely enclosed habitat — possibly a model for a lunar base. If we colonize in space or on the moon, we will need to take our habitat along with us. Volunteers entered Biosphere II with the intent to seal the doors and live there for two years, studying their problems as well as their successes.

However, there were more problems with the experiment than most people anticipated. One system malfunction caused dropping oxygen levels, and operators continued to pump oxygen into the biosphere, even after it was sealed. Medical emergencies required outside help. A period of cloudy skies and wintry weather limited solar power levels, which resulted in poorer food production and loss of nutrition.

While some scientists called Biosphere II a failure, those who designed and operated it don't agree. "These biospherians are doing the shakedown, the preparatory

> **"Think of it. We are traveling on a planet, revolving around the sun, in almost perfect symmetry. We are blessed with technology that would be indescribable to our forefathers. We have the wherewithal, the know-it-all, to feed everybody, clothe everybody, give every human on Earth a chance. We dwell instead on petty things. We kill each other. We build monuments to ourselves. What a waste of time...Think of it. What a chance we have."**
>
> *Buckminster Fuller*

work for the serious work to begin. They're still working out the bugs, and they're not to be blamed for that," explains Dr. Gerald Soffen, head scientist for NASA's Viking missions to Mars and advisor to Biosphere II.

If we can set up a base on Antarctica, we can probably set one up on the moon, and Biosphere II's failures may show scientists how to prepare for such a base. A project for life in space may not be too many solar revolutions away, astronomically speaking.

Looking farther ahead, might we one day capture and hollow out massive asteroids (using the insides as a valuable iron ore resource) and use them as inside-out worlds for entire colonies? And might these worlds be equipped to travel through interstellar space, in a generations-long voyage beyond our solar system? Once we learn to live in space, the outer borders fade away.

ACTIVITIES

CONSTRUCTING A SPACE STATION

Anything constructed in space needs to be strong, but lightweight. A good geometry student will tell you right off that the strongest geometric shape is a triangle. Force applied to any point in a triangle is dispersed to the other two sides. Three triangles hooked together into a pyramid make the strongest three-dimensional shape.

Now it's time to build something.

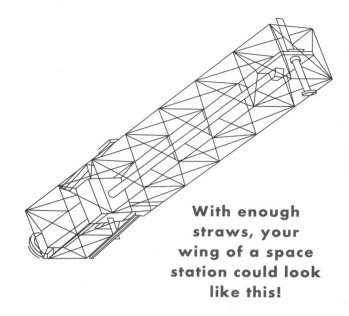

With enough straws, your wing of a space station could look like this!

> # YOU WILL NEED:
> whole bunches of straws, cut in half
>
> handfuls of pipe cleaners, cut in half or handfuls of twist ties, not cut in half

Following these illustrations, use the pipe cleaners or twist ties as connectors for the straws. Make a triangle, then construct a pyramid.

Now go for it! Use this basic model to make a space station of whatever shape and dimension you want. Will it rotate? Will there be docking ports for the shuttle? You are now the boss.

Construct.

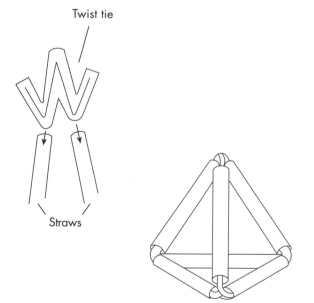

Twist tie

Straws

CROSSWORD PUZZLE OF THE STARS

The answers to this crossword come from the list of names under the heading "Service Representatives."

ACROSS

1. Dutch astronomer theorized comets' "home"

3. His name is on the "Eye of the Universe"

5. Radio waves from the sun. What a "bright" idea

7. U.S. representative for European/U.S. X-ray Multi-Mirror Mission

9. First American woman in space

11. Established system to classify star spectra

13. Discovered sun's atmosphere, mostly hydrogen and helium

15. With 17-Across, the last names in pulsars

17. See 15-Across

DOWN

2. Physicist claims, "It's all relative"

4. "Rockets Rule" slogan for past scientist

6. A Percival who observed Uranus

8. Finds galaxy where half the stars go one way and half the other way

10. Husband and wife comet watchers

12. It's all X-ray to him

14. German who mixed liquid fuel for a "rocket" good time

16. Pluto's daddy?

ACTIVITY

Feedback Data From Owners Of OMICRON IV

Your mission is to inspect and file a report on the OMICRON IV model spacecraft — called "Earth" by its inhabitants. You are to land, photograph the area, sample the lifestyle and culture of the inhabitants, retrieve samples, and file a complete report.

YOUR REPORT SHOULD INCLUDE:

photographs, with explanation and interpretation

report on culture and lifestyle

samples, with explanation of use

prediction for the future of the current dominant species

Be ready to present your report before the board of Omicron Research and Development by the next two rotations.

Spaceship Earth

One Last "Oh, Wow!" Before The Planet Whizzes Away

Actually, that's just the point. Are we "whizzing away"? If so, then in what direction and how fast?

Here's what we know about the way Spaceship Earth moves:

1. Earth rotates on its axis at a speed of 1,660 kilometers per hour at the equator (about 1,100 kilometers per hour at the middle latitudes). Direction? West to east.

2. Earth revolves around the sun at 108,000 kilometers per hour.

3. Our solar system revolves around the center of the Milky Way galaxy about once every 200,000,000 years. To do that, our sun (and us) flies through space in the direction of the star 99 Herculis at about 750,000 kilometers per hour.

4. (Hang on, it gets freaky here.) Our whole galaxy is moving (along with about 10,000 galaxies in the local supercluster) in the direction of the constellation Centaurus at about 2,000,000 kilometers per hour.

So, where *are* we going, and how *fast* are we getting there?

End Operator's Manual
All Sales Are Final

Teacher's Companion

Topics For Writing and Discussion

1. In what ways is Earth similar to a spaceship? How is it different? Is there any advantage to viewing Earth in this manner? Any disadvantage?

2. Discuss the opening quote from T. S. Eliot. How does leaving Earth help us know it better? Was this also true for explorers of the past? For astronomers of the past?

3. Does knowledge of plate tectonics have any implications for human use of our planet? How should people use this knowledge when considering such things as mining, settlement and construction, land use, etc.?

4. In our search for renewable energy sources, is there any way to "mine" the thermal and wind energy of the stratosphere and mesosphere?

5. Discuss student perceptions of the greenhouse effect as it relates to global warming. Discuss perceptions of ozone depletion, indicating future topics for research.

6. Brainstorm for other ways to help "save the Earth." Students who have read books on this subject might share what they learned.

7. What examples exist today of the use of renewable energy? What common devices use solar or other renewable energy? Discuss the use of electric cars in many major cities.

8. What questions would you be ready to ask if you were about to buy a telescope for yourself?

9. What examples are there of the use of cameras and robots to see things here on Earth that we cannot see on our own?

10. Radio waves also carry "messages" from Earth out into space. How long ago did Earth begin to send radio messages? How far into space would they have traveled and what stars would they have passed? Is this sort of "message" very effective if your goal is to try to contact extraterrestrial life? Why or why not?

11. Rocket technology was advanced quickly because of wartime applications. What other examples exist of technologies or techniques we consider positive that have developed as a result of war?

12. In a discussion of futurism, examine other trends in science and society and where they might lead in the future. Develop alternate futures from given trends.

13. (Possible research/debate topic) What are the costs of developing a lunar colony (or an orbiting colony)? What are the benefits of such a plan? Discuss/debate the costs vs. the benefits of colonization.

14. If Earth sends a team of astronauts to Mars, the journey will be very long. A principal concern is boredom. How might astronauts on a long-term mission fight boredom, and how might NASA planners prevent it?

15. Would you want to be a volunteer for a two-year experiment such as Biosphere II? What would be the benefits of being a volunteer? What would be the hardships?

16. Life has evolved and adapted to the special ecosystems of the Earth. As we consider living in space, what effect will this new "ecosystem" have upon us? What types of organisms would evolution favor in such an environment?

ADDITIONAL ACTIVITIES

1. (Multi-disciplinary, group involvement) Environmental programs are expensive and often require a diversion of substantial national resources. People in positions of responsibility must make decisions for the communities they represent.

You are a member of a city council. A one-acre plot of land has just become available to the council. Six groups have presented proposals for the land. They propose the following:

1. build 8 homes

2. plant community gardens

3. create a playground

4. build a factory

5. build an incinerator

6. do nothing with the land

For each proposal, make a chart like the one below, listing all the effects the proposal would have on the community. Then label each effect positive (+), negative (−), or neutral (o).

Discuss your results and vote on a proposal.

2. (Supplemental to No.1) You are now on the president's Space Advisory Council. A proposal for a lunar colony has been presented with a total price tag of more than $50 billion. The money, however, could be used for other national programs. Discuss other uses for the $50 billion until you come up with three plans other than the lunar colony.

Follow the same procedure as in the prior activity to analyze the four proposals. Vote on how the money will be spent.

3. (In pairs and independent) How does weightlessness feel? Some activities imitate weightlessness. Diving off a diving board provides a momentary sense of weightlessness. Swinging on a swing provides the same moment of weightlessness at the peak of the arc.

Riding an elevator also can give you a sense of a change in weight — if not weightlessness. Try this. You will need to take a few rides on an elevator while standing on a bathroom scale.

Get on the elevator and stand on the scale. Note your exact weight. As the elevator descends, what happens to your weight? What happens when you reach the bottom? What happens as you go back up? Describe your results and offer reasons for your findings.

PROPOSAL HERE		
LIST EFFECTS HERE	**+** **−**	**COMMENTS HERE**
1		
2		
3		
4		
5		
6		
7		
8		

Activity courtesy NASA publications

4. (In pairs) How can astronauts eat and drink in space if there is no gravity? How does the food fall into their stomachs?

YOU WILL NEED:

a chair

a glass of water

a straw

a piece of apple or banana

Lean over the chair so that your head is lower than your stomach. Take a sip of water. Have a bite of apple. Clearly, you can eat without gravity. Why? If gravity doesn't make food fall to your stomach, what does?

5. (Individual) Design a spectroscope.

YOU WILL NEED:

a shoe box

black construction paper

tape

diffraction grating (available from hobby or science catalogs)

Cut a hole approximately 1 centimeter square at each end of the shoe box. Line the inside of the box with black construction paper, making sure that the two holes are still clear.

Tape two pieces of cardboard on the outside of the box over one of the holes. Cover all but a narrow slit of the hole.

Tape a piece of diffraction grating on the outside of the box over the other hole. Line the box lid with black paper and tape it on the top of the shoe box.

Looking through the hole at the end with the diffraction grating, point the box at a light source. (DO NOT POINT THE BOX TOWARD THE SUN.) You will see the spectrographic "signature" of that light source. Try incandescent bulbs, candles, neon lights, and fluorescent bulbs, noting the different spectra.

6. (Individual or cooperative, homework) Often, people claim that the sun seems larger as it is setting (or rising) and that a full moon is also larger near the horizon. Some claim that the "lensing" of dust and water vapor in the atmosphere enlarges the sun and moon. Some claim this is just an optical illusion.

On the night of the next full moon, walk outside just at sunset (the full moon will be rising in the east). Hold a dime at arm's length so that the dime covers or is right next to the moon. Note how much of the moon is covered by the coin. Return outside in three hours when the moon is higher in the sky and try the same experiment. Note your results.

The next time the class meets, discuss results and brainstorm for other reasons why the full moon might seem larger near the horizon. If it is an illusion, what might be causing it?

7. (Individual or cooperative, homework or evening class activity) There are currently 400 functioning satellites in orbit about the Earth and 1,600 more that are no longer functioning. Most of these are in high orbit (33,000 kilometers) and cannot be seen without a telescope. However, many are in low orbit (500 kilometers or less) and can be seen on a clear night, especially with binoculars.

Since satellites are only visible if they reflect sunlight, low-orbit satellites must be viewed just after sundown when, even though the sun has set below the horizon for the observer, sunlight still reaches the satellite.

Look for points of light that move across the sky in 2 to 3 minutes. There should be no sounds or flashing lights associated with the observation. If one is sighted, look carefully through the binoculars. Does the satellite seem to "twinkle"? If so, it is because the satellite is rolling in its orbit, and the reflected sunlight seems to twinkle. Low-orbit polar satellites, used for mapping, are easier to spot, since they move from south to north against the sky.

You can also call *Sky & Telescope*'s "Skyline" at 617-497-4168 for a current list of observable satellites.

8. (Creative, interdisciplinary) Consider the following poem:

> *My first is in moat, but not in boat*
> *My second is in say, but not in quote*
> *My third is in gain, but not in lose*
> *My fourth is in mine, but not in youse*
> *My fifth is in ask, and also in plea*
> *My sixth has heat no sun can be*
> *The secret lies within, you see.*

Can you figure out what glossary word is the subject of the poem? "MAGMA." See how the poem works? My first (letter M) is in "moat," but not in "boat."

Create a word puzzle or poem of your own for one of the glossary words (or perhaps for a last name from the "Service Representatives" section). Each line refers to the letters of the word, and the last line offers a clue.

Bogglers' Solutions

THE COAT-HANGER PENDULUM
(page 84)

The secret to understanding this proof is to turn it around and view it from an opposite perspective. If turning the coat hanger does not change the swing of the pendulum, then a free-hanging pendulum should swing in a straight path for an indefinite period.

However, a pendulum on Earth seems to swing slowly in a circle, completing the circle in 24 hours. We know the pendulum is really swinging in a straight line, so it must be that *we* are the ones moving.

THE FIRST ROCK GROUP
(page 85)

1. lithosphere

2. magma

3. plate tectonics

4. subduction

5. jet stream

6. billion

7. radiation

8. Coriolis effect

Scrambled letters:
L,H,E,G,L,T,T,S,N,S,R,O,I,N,O,E,
Rock group: THE ROLLING STONES

BOTTLE BOGGLE
(page 89)

Answer: 111

HUBBLE CONSTANT
(page 95)

A star cluster 16 billion light-years away would be 4,908 megaparsecs away, so D = 4,908. If H = 80, then V = 392,640 kilometers per second. However, nothing can move faster than 300,000 kilometers per second, so there is a contradiction. Either the Hubble Constant is wrong (latest evidence seems to support the figure of approximately 80) or our estimates of the ages of ancient star clusters is wrong. If the Hubble Constant is 80, the universe cannot be older than 12 billion years.

PIONEER 10 PLAQUE
(page 105)

The figure of the man and woman are the most recognizable. The man's hand is raised in a gesture of greeting. The bottom shows our sun and planets, with a craft leaving the third planet. The craft is also represented as background to the two people. The "starburst" in the center represents our galaxy, with lines directing attention to known pulsars — recognizable locators for an advanced culture. The two circles at the top represent a hydrogen atom.

DOUBLE TROUBLE
(page 105)

1. Hubble stubble
2. Oort port
3. rover rover
4. pocket rocket

5. with a quasar razor
6. lunar crooner
7. Pluto kudos
8. heliopause clause
9. infrared head
10. infect-a spectra

NOTE: Anyone seriously considering the purchase of a telescope should consult the booklet *Buying Your First Telescope*, which is available from *Astronomy* magazine, Box 1612, Waukesha, WI 53187. The cost is one dollar.

CROSSWORD PUZZLE OF THE STARS
(page 115)

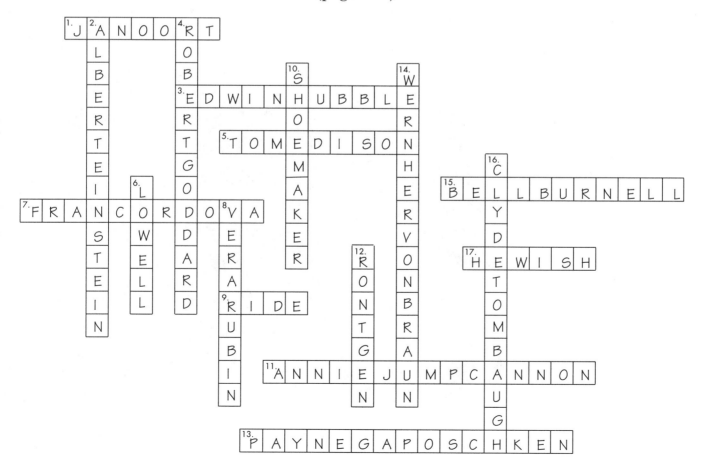

Section 4

Let The Drama Begin!

Cosmology and
Questions of the Universe

THE TERMINATOR

Relax, this is the last terminator page.

cosmology
nebula
fusion
nova
neutron
pulsar
event horizon
escape velocity
local group
red shift
dark matter
wormhole
quasar

THE AGE OF NEWTON HAS ENDED

Did you think Kepler was weird? What if Kepler was present at the time of the Big Bang? Hey, you were there, too! Did the idea of an Oort Cloud blow your mind? What about a cloud of gases swirling around the event horizon of a black hole? Was an Operations Manual for Spaceship Earth pretty far-fetched? Just wait until you see the wormhole that pulls space into a loop and brings the opposite end of the universe practically into your back yard.

The life and death of stars, the devouring of entire galaxies, the fabric of space and time itself — COSMOLOGY is the study of the cosmos, the universe, the everything. And if the rest of astronomy sometimes sounds like a soap opera, cosmology is the grandest of grand drama.

ENTER ALBERT EINSTEIN

Albert Einstein, born in Germany in 1879, didn't look like the type of student who would ever amount to much. He hated authority and rebelled against his teachers — a habit that continued even into college. ("To punish me for my contempt for authority, fate has made me an authority myself," he later noted.)

But in 1905, everything began to click for Einstein. He published four papers that year (he was only 26 years old!) that changed the way scientists thought about the world. He began the modern understanding of how the universe works at the largest level (the effects of gravity over infinite distances among galaxies) and of how the universe works at the smallest level (the behavior of particles that make up atoms, known as the science of quantum mechanics).

Einstein can be said to have invented cosmology, and his theories changed all the rules of the game. But there was one thing that always eluded him, and that still eludes cosmologists today. There really should be a way to combine gravity and quantum mechanics. Einstein never found it.

BURP

Hydrogen is the most common element in the universe — even for living things. Out of every 100 atoms in our bodies, about 50 are hydrogen.

123

Stars Are Born And Stars Die

That's right, stars have a birthday (maybe a birth epoch would be more like it). And it's also true that they cash it in just like the rest of us (well, not exactly like the rest of us). Let's peek behind the scenes at a stellar nursery and see if we can catch that happy moment....

We are here in a vast NEBULA — a swirling sea of gas and dust. Where did all this stuff come from and why doesn't anybody clean up around here? It might be that this nebula has remained since the early days of the universe, or it could be that this is the dust of an earlier generation of stars.

But look over there. Small quantities of dust and gas have collected into a ball and are sweeping a path through the nebula — the same as rolling a snowball and collecting the wet snow together as you roll. These balls of gas and dust are growing, but they aren't stars. Not yet.

To the left you can see one of the dust balls, which has become so large it is starting to contract from the force of its own gravity. Don't pass out the cigars just yet, because this is a PROTOSTAR. It may be humongous (100 times the size of the sun) but it isn't "born" just yet. The protostar begins to heat up as it is compressed. Eventually, when it has condensed to nearly the size of our sun, it becomes so hot that its radiated heat enters the infrared frequency. It becomes an INFRARED STAR.

We're almost there. The infrared star continues to collect dust and gas and continues to condense. Its core temperature is rising (and you better stand back). When the core temperature reaches 28 million degrees Kelvin, hydrogen gas (which makes up most of this baby) is transformed into helium, and nuclear fusion ignites the star's furnace.

Nuclear Reaction

5.

Red Giant

JETT·I·SON

Pulsar

White Dwarf

7.

Life Cycle of a Star

Happy Birthday!

If this infant star is anything like our own yellow star (and most of them are), it will hang around for about 10 billion years shining with the energy of hydrogen fusion. The star remains a stable size, balancing the force of its internal fusion with its gravitational force.

YOU CAN FIND A CARD AT YOUR LOCAL DRUG STORE

Our sun was born about 5 billion years ago. Party down!

A C T I V I T Y

BIRTH OF A STAR: BALLOON TRICK NO. 1

YOU WILL NEED:

a balloon (that's it)

Blow the balloon up and hold the end closed between your fingers. You have a pretty good model of a star. There are two forces that determine the size of your balloon. The air pressure inside the balloon pushes out — like the thermonuclear force inside the star. The tension of the rubber skin pulls the balloon together — like the gravity of the star.

What happens if you increase the air pressure? Sure, the balloon finds a new stable size and holds it.

The size of a star, therefore, is determined by a balance between fusion and gravity.

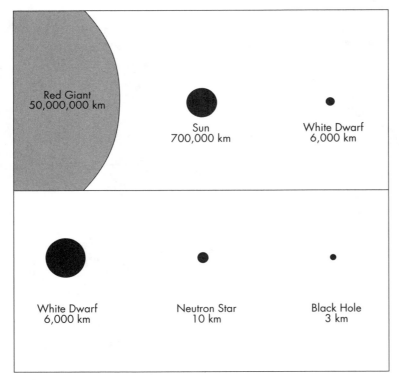

Red Giant
50,000,000 km

Sun
700,000 km

White Dwarf
6,000 km

White Dwarf
6,000 km

Neutron Star
10 km

Black Hole
3 km

Of course, not every star is like the sun. Here in the nursery you'll find some future light-weights and some future defensive linesmen. The giants and supergiants collect more dust and gas in their formative years. Are they hotter? Sometimes. The best way to guess the relative heat of a star (they're *all* hot!) is to see its color. Red stars, whether small or giant, are cooler than our sun, and blue and blue-white stars are hotter.

Life expectancy goes hand-in-corona with color, since the hottest burn themselves out the fastest. The blue supergiants are gone almost before you get to know them, while the cool, reddish stars last longer than a final exam in English class.

For the ultimate in coolness, check out a BROWN DWARF. Brown dwarf stars did not collect enough gas and dust from the nebula to emerge from the protostar phase and just sort of fizzled out. There are probably gazillions of them out there, but they're dark so no one can find them— be careful, you could stub your toe.

There are twins born in the nursery, too. BINARY or DOUBLE stars are locked in orbit around each other. Usually, the smaller star will orbit the larger one; but sometimes the two do intricate dances as each orbits the other.

But nothing lasts forever, and that goes for hydrogen as well. When a star begins to run out of hydrogen, it's like letting some air out of that balloon. The star contracts. This, however, is where the balloon image ends, because as the star contracts, its gravitational force is increasing (gravity is related to mass and density, and this fella is becoming more dense). Also, as the core temperature gets hotter, carbon and nitrogen in the star undergo thermonuclear fusion.

At this point, the outer layers separate from the core of the star and expand, forming a RED GIANT. The

WOULD YOU CARE FOR A TEASPOON OF DWARF WITH YOUR TEA?

White dwarf stars are so densely compact, one teaspoon of white dwarf stuff would weigh approximately a ton.

core of the star eventually becomes so dense it draws the outer layers back in. The star may become unstable and begin to pulsate. If you were a star, this would be a good time to get your papers in order.

DON'T BOTHER LOOKING UNDER YOUR BED

There are no black dwarf stars in the universe. A star would have to be about 25 billion years old to become a black dwarf — and the universe isn't that old yet!

The form of death for a star depends on its mass. Most stars, like our sun, continue to shrink until they are about the size of a planet. The star will spend the next several billion years as a WHITE DWARF. Eventually, all fuel is used up and the star becomes a cold BLACK DWARF — and that's about all there is, the fat lady has sung.

If the original star's mass is 1 1/2 times that of our sun, it will crush in upon itself and then explode in a NOVA. Novas throw off matter, and the star may then settle into a more typical death, as described before.

SUPERNOVAS OF THE RICH AND FAMOUS

One of the most famous supernovas ever witnessed was in 1054. It was recorded by the Chinese, the Japanese, the Turks, and American Indians.

And what has happened since? That supernova is now the Crab Nebula (in the constellation Taurus). This nebula is expanding more than 50 million miles every day.

Even more massive stars go out with a bang. A SUPERNOVA is a rare event, but it's a spectacular one. The star explodes violently, throwing dust and gas off into space. (That dust and gas, by the way, becomes a nebula — and here we go again.)

What's left after a supernova? The core of the star has collapsed to only about 10 kilometers in diameter and is so dense that only tightly packed neutrons can exist in stable form. This is a NEUTRON STAR, and if it begins to rotate and send flashes of energy into space, it is called a PULSAR.

So stars *do* die. Is that bad news? Is it sad? Life on Earth will certainly be obliterated by our sun's death — but that's billions of years away. Heck, *Beakman's World* will probably be off the air by then.

But it's not a bad thing that stars die. Look at star-death as a sort of galactic recycling project. We are all made of matter, and our matter has to come from someplace. How far back can you trace *your* matter? Mom and dad? Okay, for a start. Do you have a 300-year-old family tree? That's nice — but what about 1 billion years ago? Where was your matter 4.5 billion years ago when the Earth was just forming?

Let's face it, much of the matter that makes us must have been in that swirling nebula 4.5 to 5 billion years ago. And where did that nebula come from? Probably the death of an earlier star.

We've been recycled.

A C T I V I T Y
LOCATING FAMOUS STARS

Are you a first-timer in the field of stargazing? Do more experienced watchers laugh and kick hydrogen in your face? Here's how to catch up quickly.

YOU WILL NEED:

the following two charts
a clear night

Most people can either find the Big Dipper and Orion or see them easily if they are pointed out. If you base your stargazing on one of these familiar constellations, you can see more than a dozen "famous" stars (just like those cheesy maps to the homes of the stars in Hollywood).

B L A C K H O L E S :
T H E U N I V E R S E' S
W A Y O F
S A Y I N G
"G I M M I E"

There is one other form of star-death, but it is saved for a few special cases. The most massive stars, after they supernova, may be left with cores so dense that their gravitational force causes them to continue to contract.

Gravity is related to mass and density. As the star core shrinks, it becomes more dense. As it becomes more dense, its gravitational force becomes stronger. As its gravitational force becomes stronger, it shrinks.

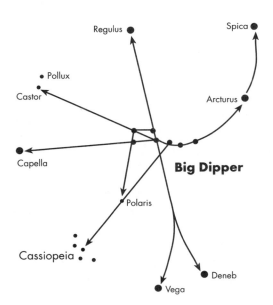

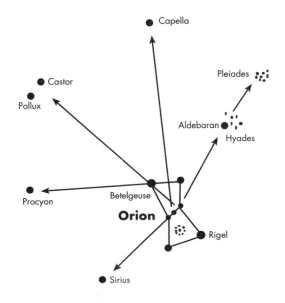

Whoa! The smaller it gets, the greater the force making it small. Nasty chain reaction.

While that star core is shrinking out of sight, here's another piece of the picture: ESCAPE VELOCITY is the speed you need to break free of an object's gravitational pull. (You need to get up to about 30,000 kilometers per hour to escape Earth's gravity, for example, and about 35,000 kilometers per hour to get out of bed on a Saturday.)

Well, guess what's happening to the speed you would need to escape from this star as it shrinks? Right, escape velocity is going way up.

So what happens when the star becomes so small, and its gravitational force so strong, that even something moving at 300,000 kilometers per second isn't going fast enough to escape? (A little *Jeopardy* music here.)

Exacta-mundo! Light isn't moving fast enough to escape this star. Light is drawn *into* this star — and since nothing moves faster than light, everything else is drawn in, too.

This is a BLACK HOLE.

What happens to things that are drawn into a black hole, and where do they go?

Good questions. Anybody want to volunteer on a mission to find out? If you were in a rocket ship approaching a black hole, you would feel the pull of gravity on you while you were still far from the black hole itself. Let's say the approach doesn't kill you (although it probably would). What would happen? You would, of course, see nothing ahead of you.

As you crossed the event horizon (the point where escape velocity is greater than the speed of light), you would no longer be able to radio for help, since radio waves would be drawn in with you. Soon, the pull of gravity on the front of your body would be so much greater than the pull on the back of your body that you would be ripped apart. As a matter of fact, each tiny piece of you would be ripped apart. (Some scientists think you would be pulled to a string-like thinness rather than ripped apart — but it seems like a minor point.) It would be

hard to go through something like that and keep your sense of humor.

Are you one of those people who wants to go through a black hole as a way to travel through time? Either you've been watching too many movies or you don't mind if people call you spaghetti head. You don't want a black hole, you want a wormhole — but hang on a few pages.

ACTIVITIES

DEMONSTRATING ESCAPE VELOCITY (GET OUT OF TOWN)

Here's how escape velocity works:

YOU WILL NEED:

a board at least 4 feet long

a cardboard roll from wrapping paper

some marbles

a jar of peanut butter
some tape

Tape the cardboard roll lengthwise to the board, leaving at least six inches of board at one end. Scoop out a good hunk of peanut butter and spread it on the exposed end of the board so that the marble can roll out of the tube and over the peanut butter trap (which should be about one-half inch thick).

Tilt the board slightly and roll a marble down the tube. Did it get stuck in the peanut butter trap? If it did, then it didn't have enough speed to break free — it didn't reach escape velocity. Find the elevation necessary for your marble to reach escape velocity. In space, the force of gra-

vity is the "trap." In black holes, the gravitational trap is so massive even marbles going the speed of light can't escape.

Now get some crackers and eat your project.

MAKING A BLACK HOLE — SORT OF

If you want to get technical, gravity doesn't really "pull." The reason gravity seems to pull objects is that all matter causes dips and bends in space. The greater an object's mass, the more space is bent around it, and the harder it is for an object to "climb out of the hole." (What? Suddenly, you don't want to get technical?)

YOU WILL NEED:

an old T-shirt

a rubber trash can or a large bucket

an assortment of marbles and small stones

some twine

Cut up one side of the T-shirt until you reach the reinforcing material around the collar. Don't cut through that.

Spread the T-shirt over the top of the trash can (you might want to hose that can out first or this could be a stinky experiment). Make sure that the open neck is at one edge of the trash can.

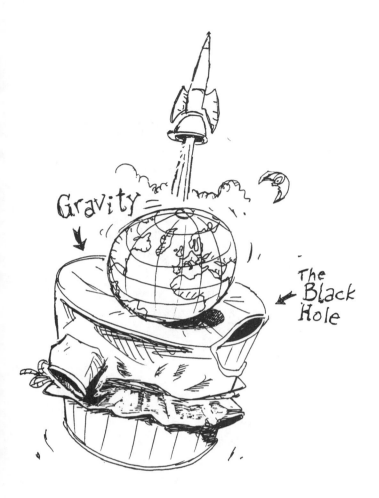

Gravity

The Black Hole

Tie the T-shirt to the open top of the trash can with the twine. (You might want to get the help of a friend to hold everything in place while you tie.) Try to get the shirt tight enough so that it looks flat like a drum head.

Your shirt represents space. Carefully place a marble on the shirt. What happens to space around the marble? Now try a heavier stone. Observe how greater mass bends space more. If you were in a tiny rocket and that rock were your home planet, you would have to get up enough speed to "climb" out of that bend in space. That would be your escape velocity.

Okay, now carefully stick your arm through the neck hole and grab your T-shirt from underneath. Pull down.

A black hole warps space so severely that nothing can go fast enough to climb out.

GALAXIES: TWO'S COMPANY, 200 BILLION IS A CROWD

Pack enough stars together, send them whirling around each other in a gravitational dance, and you've got yourself a galaxy. But don't get too uppity about it, there are hundreds of billions of them.

To the naked eye, galaxies look like stars. While galaxies may contain hundreds of billions of stars, they are so far away you need a telescope to realize they aren't just single points of light. There are three galaxies you can see without the aid of a telescope. The two best views are in the southern hemisphere, where you can see the two Megellenic Clouds, next-door-neighbor galaxies a mere 160,000 light-years away. The only galaxy visible to the naked eye in the northern hemisphere (and you need a clear sky and a good eye) is our nearest major galaxy, Andromeda (located in the Andromeda constellation).

Actually, there is one other galaxy visible to the naked eye...ours. We orbit a star in one of the spiraling arms of the Milky Way. If you look overhead on a clear night, you can see a thick band of stars stretching across the sky. That band is our arm. If you are in the southern hemisphere, you can see the center of the galaxy as well.

By the way, do you notice the dark band that runs through that bright arm of stars overhead? That's called the Great Rift. It isn't that there are fewer stars there; the Great Rift is dark because stellar dust and gases blot out many of the stars. In the past, different civilizations have described this dark band many ways — including the "backbone of the sky."

131

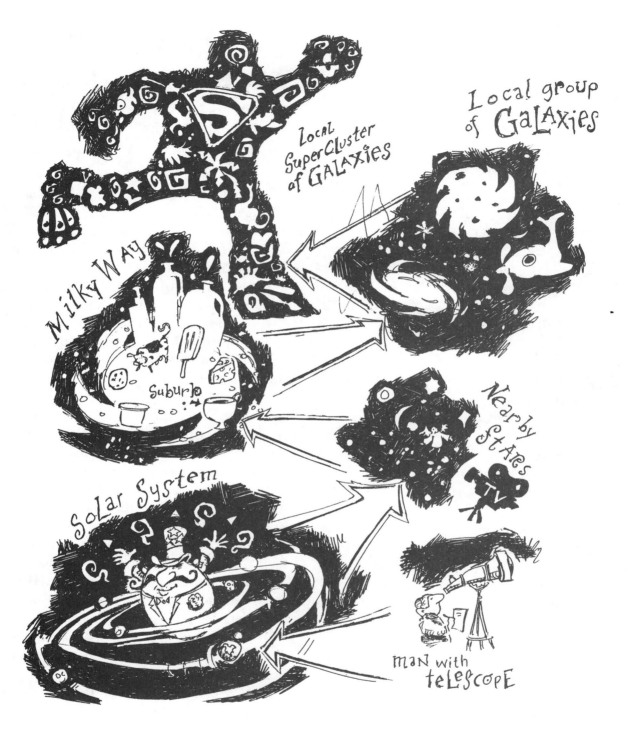

Local SuperCluster of GALAXIES

Local group of GALAXIES

Milky Way

Suburb

Nearby Stars

TV

Solar System

man with telescope

Andromeda, the Milky Way, the two Magellenic Clouds, and an assortment of smaller galaxies make up what's called the local group — kind of the neighborhood gang.

Actually, that won't always be true. The Milky Way is slowly moving toward the Magellenic Clouds and will one day gobble them up like nachos at a party. Andromeda is engaged in the same kind of intergalactic cannibalism — and one day our local group will be made up of only Andromeda and the Milky Way, each moving in to devour the other.

Galaxies, like stars and planets, come in many shapes and sizes. There are swirling spiral

NOW THAT'S PLAYING IN THE BIG LEAGUES

If our solar system were the size of a baseball, the Milky Way would be the size of North America.

galaxies, with smooth or lumpy arms circling the galactic core. The Milky Way is one of those spiral galaxies, and our sun is whirling on one of those arms.

Barred spiral galaxies twirl as if they were one solid mass, with two sharply defined arms of stars. Elliptical galaxies don't twirl. Through telescopes they seem like large, fuzzy tennis balls, as each of

the millions of stars establishes its own orbit around the galactic core. Irregular galaxies seem disorganized and may be young, since the amount of gas in them is high. The stars in irregular galaxies seem to be drawn to a central mob, just like students drawn to the odor of a hot pizza.

Stars definitely like to hang out together. But no matter where we look in the sky (through a telescope), galaxies are there — we are surrounded by them. And, with the exception of the galaxies in our local group, there is only one thing they all have in common: They are all moving away from us.

As a matter of fact, some are moving away so quickly that it makes you wonder if the Milky Way has some sort of cosmic B.O.

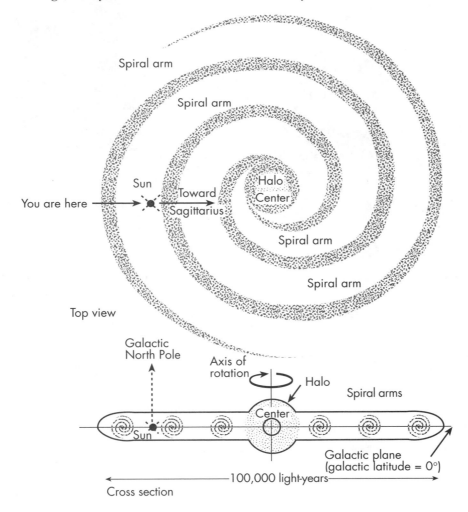

Diagram of Milky Way

ACTIVITIES

GRAPHING THE LOCAL GROUP

Would you like to know exactly where all those other galaxies are in our local group?

YOU WILL NEED:

a piece of graph paper
colored pencils

There are actually 26 galactic bodies in our local group. The Milky Way and Andromeda are by far the largest. The two Magellanic Clouds are smaller, and then there are a whole bunch of "dwarf galaxies."

Number both the vertical and horizontal axes of your graph paper from 1 to 100. Each pair of numbers on the next page will be given with the horizontal axis first, then the vertical. For example, find 38 along the horizontal axis. Now find 63 along the vertical axis. Mark the point where those two lines meet with a red pencil. That will be the location of the Milky Way. The Milky Way would be listed as (38,63). See how it works? (You might want to make a "key" as you go so that you can remember later which marks belong to which galaxies.)

Andromeda deserves a red mark, also, and you can locate it at 67,65.

A green pencil mark is reserved for the next smallest group, which includes: Large Magellanic

Cloud (43,62), Small Magellanic Cloud (41,63), and the galaxy called M33 (75,74).

The list below includes the other 21 members of our local group. They are dwarf galaxies and should be marked with a black pencil.

Milky Way group	Andromeda group
Leo I (34,67)	NGC 147 (61,65)
Carina (38,67)	M32 (62,65)
Leo II (32,63)	NGC 185 (63,64)
Sextans (35,63)	NGC 205 (65,65)
Sculptor (47,62)	A-II (69,65)
Fornax (48,66)	A-I (72,65)
Ursa Minor (35,57)	A-III (73,63)
Draco (38,54)	WLM (75,61)
Gets pretty crowded, doesn't it?	IC 1613 (82,73)
	Pisces (88,73)
	(and there are always a few "hangers on")
	NGC 6822 (54,35)
	DDO 210 (72,24)
	GR8 (0.5, 0.8)

There's the neighborhood. Hang on to that map in case you get an intergalactic paper route one day.

MOVING GALAXIES: BALLOON TRICK NO. 2

No matter where we look in the sky, galaxies (unless they are in our local group) are moving away from us. What does that mean? Is the Milky Way in the center of the universe? Probably not. There is a more likely explanation.

Blow up the balloon just enough so that you can make some marks on it with a marker.

Mark a dot that will represent our galaxy. Then make a dozen other dots to represent other galaxies.

Now, blow up the balloon completely. Imagine what you would see if you were on a planet in one of those galaxies. Right. Every other galaxy would be moving away from you. However, you are not really in the center of the universe. Your position is just like every other galaxy, and no matter which galaxy you were on, you would still see everyone else moving away.

Most scientists accept this explanation for why galaxies move apart rather than the idea that we are at the "center."

By the Way!

How do we know other galaxies are flying away from us?

You could say we know because Edwin Hubble told us. In 1925 Hubble identified a type of star called a Cepheid variable within the Andromeda galaxy. Not only did this confirm that Andromeda was a galaxy rather than a nebula (a common theory of the time), but it also made it possible to measure how far away Andromeda was.

In 1929 Hubble realized that galaxies are flying apart from each other because he examined the spectra coming to us from those galaxies. (You do things like that and people name telescopes after you.) Hubble knew the general "signature" spectra of galaxies, but found that the expected patterns were shifted to the red end of the spectrum. The phenomenon is called RED SHIFT.

So what?

Well, you wouldn't ask that if you lived near a railroad.

Did you ever stand by the railroad tracks (or wait in the car) as a train went by? Did you notice the change in the sound of the whistle as the train passed? It seems to get lower as the train goes by.

The reason is pretty simple. Sound travels in waves. High-frequency waves are high notes and low-frequency waves are low notes. The frequency of a sound wave is how quickly each wave gets to your ear. Picture the sound waves that come from a train. As the train comes toward you, those waves are "pushed ahead" and arrive faster than if the train had been standing still. As the train leaves, those waves are "dragged along." Result: As the train approaches, the whistle is higher than it should be, and as the train passes, it's lower.

Hubble proved that we are living in an expanding universe.

DEMONSTRATING DOPPLER

Give your friend a whistle and tell him to blow it. Note the sound. Now, tell your friend to back off about 20 feet and come running past you, blowing the whistle as he goes. As your friend does this, notice how the sound of the whistle changes. Be careful how many times you ask your friend to repeat this.

(If you don't have any friends who are this understanding, have a not-quite-so-understanding friend blow the whistle out of the window of a car as someone drives by you.)

This phenomenon is called the Doppler effect, and it applies to waves of light just as it does to

waves of sound. If a light source (like a star or a galaxy) is moving away from us, it's spectrographic fingerprint will be shifted to the lower end. In other words, red shift. If a galaxy were flying toward us, it would be blue shifted.

Feeling nebulous? "Nebula" means "cloud" in Latin, so get your head out of the nebulae and figure this out.

There are several kinds of nebulae. REFLECTION nebulae shine by reflecting light from nearby stars. EMISSION nebulae have enough heat and energy to shine on their own. An outer shell of a dying star is called a PLANETARY nebula. Finally, DARK nebulae are clouds of dust so thick they block the light of stars behind or within them.

Consider the following six nebulae: the Ring nebula, the Lagoon nebula, the Horsehead nebula, the Dumbbell nebula, the Trifid nebula, and the Orion nebula. You can find them in the following constellations: two are in Sagittarius, two are in Orion, one is in Lyra, and one is in Vulpecula. Not only that, three of them are emission nebulae, two are planetary nebulae, and only one is a dark nebula.

Your job is to match the nebula with its constellation and its type. How? By reading these four statements, that's how.

1. Lagoon and Trifid are from the same constellation, but neither is dark like one of the nebula from Orion.

2. The two planetary nebulae are from constellations that have only one nebula each.

3. Horsehead is dark.

4. Dumbbell and Ring are not emission nebulae, but Dumbbell's constellation appears later in the galactic index than Ring's constellation.

(And those are the only hints you need.)

THE BIG BANG: THE ULTIMATE BIRTHDAY

If the galaxies of the universe are flying apart from each other, what does that mean about where they were yesterday? Right! Yesterday, they must have been closer together. What about 100 years ago? What about 5 billion years ago?

If you think backward, there must have been a time when all the matter of the universe was gathered into a very small place. The idea of the Big Bang theory is that, sometime between 8 and 20 billion years ago, the universe began in a tremendous burst of energy. As that energy began to spread apart and cool, hydrogen and helium began to collect into large clouds. Those clouds began to collapse and form stars and galaxies and planets and everything else. Today, when we see galaxies flying apart from each other, we are seeing the continuing effects of that initial blast.

In 1992, NASA's Cosmic Background Explorer satellite (COBE) mapped the background radiation of the universe. Theoretically, this is the radiation that existed 300,000 years after the Big Bang (a mere blink of a cosmic eye). The COBE map shows how gases might have formed with just enough unevenness to begin to clump together. Keep in mind, all of this is just a theory, but it seems like a pretty good one.

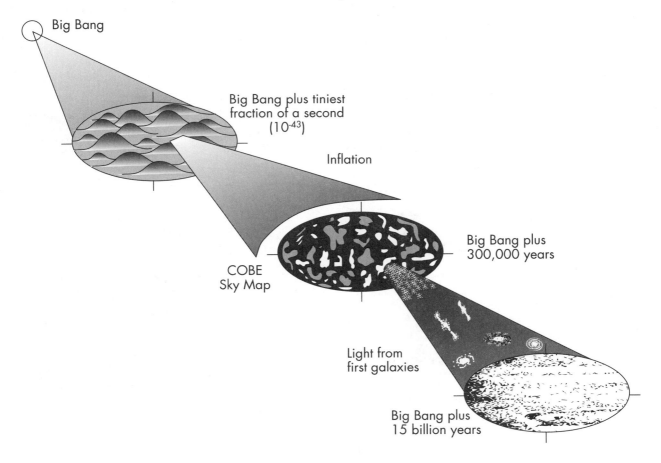

Big Bang

Big Bang plus tiniest fraction of a second (10^{-43})

Inflation

COBE Sky Map

Big Bang plus 300,000 years

Light from first galaxies

Big Bang plus 15 billion years

Big Bang Double Boggler

Happy 20-billionth birthday, universe (give or take a billion). After all this time, you still can make the news. Here to help you celebrate are many famous names involved with the Big Bang theory.

Here is a list of 10 of them: Edwin Hubble, Penzias & Wilson, George Gamow, Stephen Hawking, Alexander Friedmann, Henrietta Leavitt, Albert Einstein, Vesto Slipher, Arthur Eddington, Willem deSitter.

Some of these names you may know, but all of these people played a part in the evolution of the Big Bang theory. What follows are 10 descriptions. All you have to do is match a name to a description (the spaces for filling in the names will help you make the right match).

But wait. Did you think you would get only one puzzle on something as big as the Big Bang? No way! Here are two puzzles.

After you fill in all the names, transfer the circled letters to their correct numbered spaces below to reveal a secret message. Two puzzles for the price of one (but, of course, you already knew this book was a bargain).

1. He was the granddaddy of the expanding-universe theory after the 1917 publication of his theory of relativity.

[_] __ __ __ __ __
21

[_]__ __ __ __ __ __ [_]
16   26

2. The first observable evidence of an expanding universe came when he saw galaxies moving away from each other at high speeds.

__ __ __ __ [_]
12

__ __ __ __ [_] __ [_]
15   2

3. He puzzled over Einstein's ideas and came up with a solution that indicated a Big Bang.

__ __ __ __ [_] __
4

__ __ __ __ __ __ __ [_] __
3

4. He found a mistake in Einstein's work and discovered another theory for an expanding universe.

__ __ __ __ [_] __ __ __
25

[_] __ __ __ __ [_] __ __ __
10   13

5. Although suffering from a crippling disease, this physicist's recent work has brought Einstein's ideas (and his own) to millions.

[_] __ __ __ __ __ __
9

__ __ __ __ __ __ [_]
27

6. The Big Bang's "P.R. man," he made sure scientific ideas on the subject were presented to the public.

__ __ __ __ __ [_]
11

__ __ __ __ __ [_] __ __ __
24

7. He put ideas and observations together to make the Big Bang theory believable. We even named an orbiting telescope after him.

__ __ __ __ __ __ __ [_] [_] __ __
17 20

8. He coined the phrase "Big Bang."

[_] __ __ __ [_] __
1, 8   19

[_] __ __ __ __
23

9. Her discovery of supergiant stars helped measure distances millions of light-years away.

__ __ [_] __ __ __ [_] __ __
      7   14
__ __ __ __ __ __[_]
                5

10. While working for the telephone company, this team discovered the background radiation left behind by the Big Bang.

__ __ [_] __ [_] __ __ &
      22   18
__ [_] __ __ __ __
 6

To uncover a secret message, transfer the numbered letters above into the spaces below.

__ __ __ __ __ __ __ __ __
 1 2 3 4 5 6 7 8 9

__ __ __ __ __ __ __
10 11 12 13 14 15 16

__ __ __ - __ __ __ __
17 18 19   20 21 22 23
__ __ __ __ !
24 25 26 27

(Reprinted from ODYSSEY, December 1992, page 44.)

The Cosmic Timeline

WARNING: No one knows for sure how old the universe really is. What follows is based on a guess that the universe is about 15 billion years old.

At Time Zero, all the matter of the universe, for some reason, exploded into being. For the first tiny fraction of a second, no one has any clue about what happened. This first instant

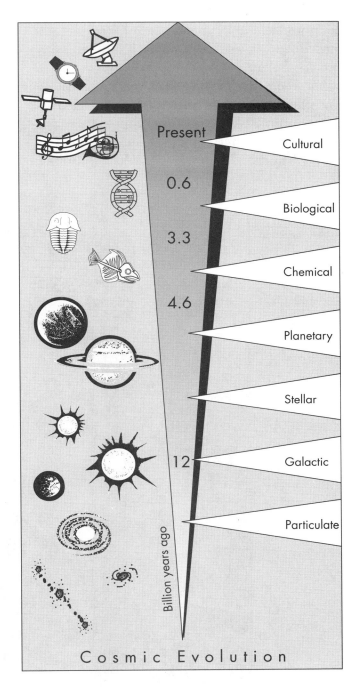

Cosmic Evolution

$(10^{-43}$ of a second) was so hot and dense that our current laws of physics don't apply. (This unknown time is called Planck time, after Max Planck, the founder of quantum mechanics.) After the first fraction of a second, things had cooled enough for quarks, the building blocks of protons, neutrons, and electrons, to form. After about 100 seconds, protons and neutrons were able to combine to form the nuclei of what would (in about a million years) become helium.

For about the first 300,000 years after the Big Bang, conditions were too intense for matter to form. At this time, the universe was all energy. Then, after perhaps a million years, things had settled down enough for electrons to orbit atomic nuclei — and hydrogen and helium were born. This was also about the time photons were able to fly around. Before this time, the universe had been a dark place, but now there was light.

The formation of hydrogen, helium, and a few other light elements took place between 1 million years and 1 billion years after the Big

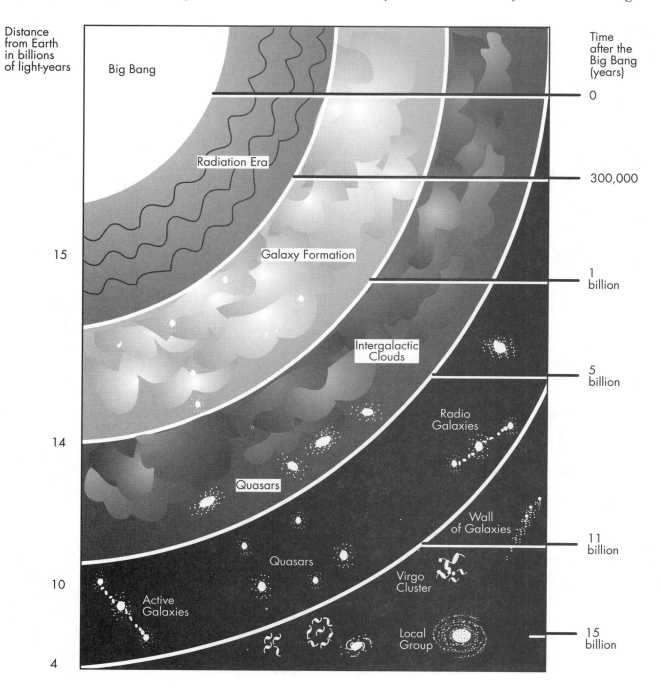

Bang. These light elements began to collect around each other, and intergalactic clouds of matter (which would later form the first proto-stars) and the first quasars appeared during the next few billion years.

The first galaxies appeared 5 to 10 billion years ago and were high-energy radio sources (radio galaxies) or active galaxies. These galaxies released much more energy than present-day galaxies, giving a hint to the violent tidal convulsions within them.

It wasn't until about 10 or 11 billion years after the Big Bang that conditions had cooled enough for the relatively stable stars and galaxies we know to exist.

The Big Bang theory is such a freaky concept that some equally freaky questions come to mind.

For example, if the universe is expanding, what is it expanding into?

Actually, it isn't expanding *into* anything. There isn't a whole bunch of space out there waiting for the universe to grow into it. Space *is* the universe! As the universe expands, it expands space as well.

GIVE ME A STATUS REPORT, COMMANDER DATA

How big is the universe at this point in its history?

There's some heavy guess work here, but one good estimate of its radius is 167 billion trillion kilometers (167 with 21 zeros). Of course, while you read that little tidbit, the radius grew by about 500,000 kilometers.

Okay, then, if the Big Bang is how things began, how will they end?

You have three choices here, and what eventually happens to the universe all depends on how much matter there is in it. If the universe is expanding because of the Big Bang, what about the force of gravity? Gravity would tend to make things collapse, not expand. Gravity, however, is not a very strong force, and, so far, the contest between the explosive force of the Big Bang and the contracting force of gravity has been pretty one-sided.

If there is enough matter in the universe, then there will eventually be enough gravity to overcome the expansion of the universe. If there isn't enough matter, then things will expand forever.

Choice No. 1. There Is Enough Matter

The universe will, hundreds of billions of years from now, slow to a halt. At that point, gravity will take over and everything will begin to contract. As matter gets closer together, it will contract faster. What will eventually happen? Well, it might be that all the matter of the universe will come crushing in upon itself into a Big Crunch! And then? Bang! This is called the Oscillating theory, and it describes a universe that goes through periodic expansion and contraction — like some massive, breathing beast. It could be, however, that the collapse of our universe does not create a second Big Bang, and all of this only happens once.

Choice No. 2. There Is Just Enough Matter

This is called the Steady-State theory, and it describes a universe where expansion slowly comes to a halt, but not until matter is so spread apart that gravity does not cause it to collapse. In such a universe, all matter would eventually decay to energy and all energy would eventually dissipate through space over immense distances.

Choice No. 3. There Is Not Enough Matter

According to the Expansion theory, galaxies may slow down, but they will never stop racing away from each other. Eventually, each galaxy will be alone in distant space. Like the Steady-State theory, matter would be recycled, but would eventually decay and dissipate — and the universe would end in cold nothingness.

ACTIVITY
GRAVITY HANGS IN THERE

Throw a baseball. Throw it as hard as you can. You can make it sail for a long time, if you have a good arm. The distance the baseball flies is the result of how much force you put into your throw.

But which force wins? Right, gravity. Gravity may not be the most powerful force in the universe, but it hangs in there.

DARK MATTER

So is there enough matter? Not according to what scientists have discovered so far. Adding up all the known matter of the universe (stars, nebulae, etc.) only gives about one-tenth of the matter needed to halt expansion.

Of course, they haven't counted the DARK MATTER yet.

You might remember that astronomers discovered Neptune not by seeing it, but because something was having a gravitational effect on Uranus' orbit. (You *do* remember that, don't you?) Well, the same might be said for galaxies. Close inspection of stars within galaxies and how galaxies relate to each other shows that there is something else out there. Dark matter may consist of planets or brown dwarfs or dust or even sub-atomic particles called neutrinos.

The point is, estimates of the visible mass of galaxies don't add up to their gravitational effects. On the other hand, early findings from the Hubble Space Telescope seem to show there isn't as much dark matter as we thought. Pretty mysterious, huh? We don't know what, and we don't know how much, but something *is* out there.

There is no way yet of measuring how much dark matter exists in the universe. Not too long ago, most scientists didn't believe it amounted to much. However, current thinking is that dark matter may actually account for 90 percent of the universe. The bottom line is that it's going to be a close call among the three theories of how the universe will end.

MAYBE THERE'S A CHOCOLATE MILKY WAY

Another theory of dark matter involves entire "dark galaxies." These are loose collections of stars that are not close enough together to be seen by even our best telescopes. (Remember that even large galaxies are not very easily seen.)

ACTIVITY

DIAL THE EVOLUTION OF THE UNIVERSE

You can create a visual dial to show the evolution of the universe.

> **YOU WILL NEED:**
> a 5.25-inch computer disk
> scissors
> thin cardboard
> colored pencils
> a ruler

Cut about one-quarter inch off the end of the computer disk on the side opposite the oblong "reading" hole. (If you don't get along with computers, this is the best part.) Remove the round diskette.

Trace the diskette on a piece of thin cardboard and cut the circle.

Use the scissors to enlarge the reading hole so that it is about the same size as the central hole. Then cut a V-shaped notch on the side that you have already cut open. Enough with the scissors, already.

Use the ruler to divide the cardboard circle into six "pie slices." (Don't worry if they are not exactly the same size.)

Each of these slices will represent a period in the history of the universe, and your job is to draw a picture representing each period in the outer inch of each slice (see "Slices" diagram).

The six periods are:

1. The Big Bang (time zero)

2. The Radiation Era (time 300,000 years)

3. The Era of Galaxy Formation (time 1 billion years)

4. The Era of Quasars and Intergalactic Clouds (time 5 billion years)

5. The Era of Quasars and Radio Galaxies (time 11 billion years)

6. Today

What should these pictures look like? Use your imagination.

At the very inside "tip" of each pie slice, label the era (and the time, if your penmanship is tiny enough).

Insert the history disk into the disk cover. Turn the disk at the V-shaped cut you made earlier. Your pictures should, when turned, show the evolution of our universe.

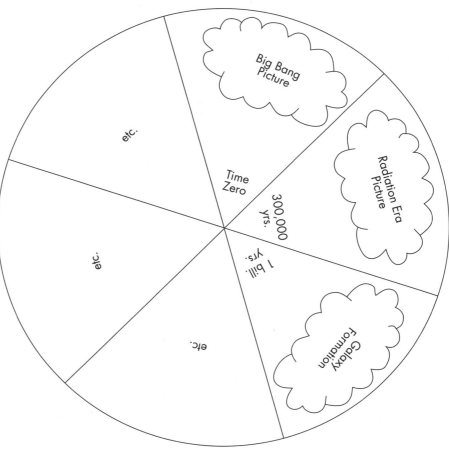

Time: Can I Get There Yesterday? And If I Can, Will My Book Report Still Be Late?

What is time? Is it a real thing or just something we made up to keep ourselves on schedule? Why do we remember the past and not the future? Is time travel possible? Will these questions go on until the end of the book?

Albert Einstein, and the physicists who followed him, have shown us that time is a dimension — like the three dimensions we use in geometry. That makes sense. After all, to locate an object in space, you need three coordinates: the object's depth, its width, and its height. But what if the object moves? To precisely locate something you need a fourth dimension — time. Now you can locate not only where the object is, but when it will be there. Cosmologists refer to these four dimensions as space-time.

But we can move through the other three dimensions easily. Why not time?

The reason is that time is moving also. Time moves. Think of the sun. You know that the sun is about eight light-minutes from Earth. That means that, when we look at the sun (even though we shouldn't do that), we are not seeing the sun now, we are seeing it eight minutes ago. We are looking eight minutes into the past.

> It would be nice to be able to travel into the future and come back and tell everyone how things will work out. But then again, if that were possible, wouldn't someone from the future have already come back to tell us? Hmmm....

You cannot separate time and light, and some scientists believe they are actually two parts of the same thing.

Let's say, for example, you were orbiting the sun when it suddenly blacked out. You realize that, back on Earth, the sun is still shining — but in eight minutes they are in for one heck of a shock. If you were able to get to Earth to warn people in, let's say five minutes, they would have three extra minutes to prepare themselves — at least enough time to find a flashlight.

Now you know the light from the sun travels 300,000 kilometers per second on its way to Earth, so you know that you would have to travel faster than that to beat the blackout. Let's say you somehow go at nearly twice that speed — 500,000 kilometers per second. You would reach Earth in about five minutes.

So you reach Earth and warn all people to find their flashlights. There are still two minutes left. Where are you? You look at the sun and notice that it is still shining. You remember that you were orbiting the sun when it was still shining. And yet, you're not there, you're here. Apparently, you have traveled back into Earth's

---

TIME TRAVEL MADE SIMPLE

Would you like to journey into the past? Nothing could be easier. All you need to do is look into the sky on a starry night. What you see is light that came from the stars long ago and has spent the intervening years heading this way to your eyes. If you see the North Star (Polaris), you are looking at light that left that star 780 years ago. You are looking 780 years into the past.

THIS WILL TELL YOU MORE ABOUT YOU THAN IT WILL ABOUT TIME, BUT TRY IT ANYWAY

past, even though you have aged. Conclusion: If you travel faster than the speed of light, you can go back in time.

Think carefully and answer: If you were given a one-way trip on a time machine, and had to take it, where and when would you go to, and why?

If only it were that simple. The problem, as Einstein first noted, is that the faster you move through space, the greater your mass becomes, and the greater your mass, the greater the force needed to make you accelerate further. As you approach the speed of light, your mass climbs toward infinity, and the energy required to speed you up further also approaches infinity.

Nothing can go faster than the speed of light.

Nobody understands time better than cosmologist supreme, Stephen W. Hawking, whose book *A Brief History of Time* has become *the* word on such things. Hawking describes why time works the way it does — in one direction only. He calls it the Arrow of Time. Time moves forward in three ways:

First, there is the Thermodynamic Arrow. This describes how things become more "disordered" as time passes. Energy is used up rather than created, even at the level of the atom.

Second, there is the Psychological Arrow. Simply, we remember yesterday, not tomorrow. We *feel* time pass this way.

Third, there is the Cosmological Arrow, showing how the universe is expanding rather than contracting.

Put these three arrows together and you have an unmistakable direction of time. You can't change time's direction, and that's that.

...But you can trick it.

WORMHOLES: WHERE SPACE EQUALS TIME

What is a wormhole, anyway?

Do you remember that T-shirt that you mutilated in order to show how matter bends space? Well, picture what would happen if you cut the bottom off that trash can and put another T-shirt over the bottom. We will let the top shirt be the space we call the Milky Way. The bottom will be the space we call Andromeda. The side of the trash can is the impossible distance you would have to travel to go from the Milky Way to Andromeda (more than 2 million light-years).

Now, take a hold of the underside of both T-shirts and pull them together inside the trash can. That is a wormhole.

Wormholes are cosmic shortcuts. If you could build a wormhole to take you from the Milky Way to Andromeda in five years, you could arrive faster than light.

Are such things possible? Well, there is nothing in our knowledge of physics that says they aren't possible. Wormholes probably would not exist naturally, however, so we would have to build one — and building a wormhole would

require much greater technology than we possess...yet.

SO WHERE DO WE GO FROM HERE?

The neat thing about astronomy is that, with every discovery, the questions keep getting better. The more we figure out about the universe, the more we realize we still have to learn.

For example, there are "things" out there we haven't got a handle on yet — like QUASARS.

At first, when astronomers discovered these faint, bluish objects, they thought they were a kind of star. Unlike stars, however, quasars emit most of their energy at infrared and radio frequencies. They are billions of light-years away, yet provide the strongest infrared signals. Quasar 3C273 is 3 billion light-years away. Even though most of its energy is invisible infrared, it is still the brightest object ever observed. Because of its distance, you can't see 3C273 without a telescope, yet it is 5 trillion times as bright as our sun.[*] Quasars might be explosive cores of galaxies. But whatever they

> **WOULD THERE BE A MARKET FOR SUNSCREEN FACTOR NO. 884?**
>
> If quasar 3C273 were as close to the sun as Alpha Centauri (4.3 light-years) it would shine on Earth with the brightness of 68 suns.[*]

[*] McAleer, Neil, page 144.

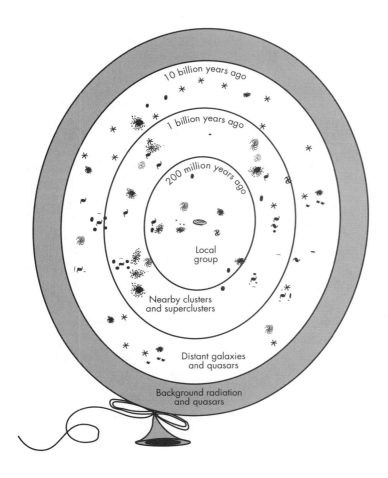

10 billion years ago

1 billion years ago

200 million years ago

Local
group

Nearby clusters
and superclusters

Distant galaxies
and quasars

Background radiation
and quasars

are, they are certainly among the weirdest things in space.

The great mystery for astrophysics is the search for a theory that will explain how the universe works. Einstein's theory of relativity explains how large things work — stars and galaxies and gravity. Quantum mechanics explains how small things work — quarks and other subatomic particles.

But there isn't a way to combine these two general theories into a THEORY OF EVERY-THING (TOE). Why does anyone look for such a theory?

Our science can explain the universe to a point. That point is one ten-millionth-trillionth-trillionth-trillionth of a second after the Big Bang (Planck time, as mentioned earlier). Before this, our scientific laws don't work. That must mean that we haven't found all the right laws yet. Result: a quest for TOE.

And speaking of quests, how about the quest for extraterrestrial company? Is life unique to Earth, or is it a common occurrence in the universe? And if anyone else is out there, can we ever make contact?

So where are they?

Project SETI (the Search for ExtraTerrestrial Intelligence) and its descendent, Project META (Mega-channel ExtraTerrestrial Array), have come up empty-handed. Although astronomers have scanned the entire sky for radio signals at 21 centimeters (the natural frequency of hydrogen — a logical broadcasting choice for anyone seeking contact in a universe that is mostly hydrogen), they haven't found anything conclusive.

Why not?

There are many reasons why such a search would be negative. Maybe we have evolved intelligence earlier than anyone else, and we are looking for radio signals from life forms that haven't developed radio yet.

Maybe other civilizations are much farther advanced, and no longer use radio signals. These civilizations may contact each other frequently, using technology we haven't yet developed.

Maybe we are being deliberately avoided. (That sounds a little creepy.)

Maybe we need to be patient. After all, it's a big universe.

Maybe we've been unlucky.

Or, maybe we're alone.

If the day comes when we make contact with another technologically advanced species, it will change our world like nothing has before. On the other hand, if the day ever comes when we discover that we *are* alone — that our planet is the *only* one among trillions of planets given the gift of life — and that we are the *only* beings in the universe ever to have established a

147

technological civilization — that will also change our world like nothing ever has before.

Even the builders of Stonehenge would agree: Astronomy doesn't get any better than this.

ACTIVITY

HOW MANY E.T.s ARE OUT THERE?

How can you determine how many civilizations might be out there for us to contact?

YOU WILL NEED:
a calculator

paper

Before we start thinking about extraterrestrial contact, we should have an idea of what our chances are. Scientists at Cornell University developed the following step-by-step approach to figuring out how many civilizations might exist for contact within the Milky Way. (Any advanced civilization outside the Milky Way would be too far away to consider.)

DISCLAIMER: This is all based on guesswork. Educated guesswork, but guesswork just the same. Also, not available in Wisconsin. Some prices may vary.

1. Life must exist on a planet around a star. So how many stars are there in the Milky Way? (Educated guess: about 200 billion.)

2. Of those stars, how many would support a planetary system? (Educated guess: about 1 in 4.) That gives us 50 billion planetary systems.

3. How many planets in a planetary system are capable of supporting organic life? (Guess

based on our solar system: 2 per system.) That gives us 100 billion planets.

4. What are the chances of life originating on a habitable planet? (Guess based on our solar system: 1 out of 2.) That gives us 50 billion planets where life has begun.

5. What are the chances life will survive to develop intelligence? (Who knows, so we'll be conservative and say 1 out of 10.) That gives us 5 billion planets with intelligent life.

6. What are the chances that intelligent life will develop the technology necessary for contact? (Same conservative estimate as above, 1 in 10.) That gives us 500 million planets in the Milky Way with technologically advanced civilizations.

But wait. There is one nasty piece of the formula that we have to include, even though we have absolutely no way to estimate it. When a civilization develops the technology to make contact, they also have developed the technology to destroy themselves. So...

7. What are the chances that a technologically advanced civilization has destroyed itself (either through violence or harm to the environment, for example) and is therefore *not* available for contact (or anything else, for that matter)?

We can't estimate this last piece of the formula, because we have no experience. True, we have survived our own technology. But to be honest, it's only been about 40 years since we've been able to wipe ourselves out — that's hardly a record we can count on yet. The best way to factor in this piece is to try different numbers. Let's be gloomy first. Let's say that out of every 1,000 advanced civilizations, only one survives its own technology. That would leave 500,000 advanced civilizations in the Milky Way ready for contact. Try some different numbers in this last piece, and see what results they give.

Any way you look at it, the possibilities do seem pretty good.

Teacher's Companion

Topics For Writing and Discussion

1. There are many myths about Albert Einstein. Make a list of all the things you think you know about Einstein. Then check to see if any of them are not true.

2. We visited a nursery for stars. What would a planetary nursery look like? What would you expect to happen there?

3. From what you know of the colors of the spectrum, why do you think the hottest stars are blue-white and the coolest stars are red?

4. What would life be like in a planetary system with a double star? How might life evolve differently, and what would daily life be like if we lived on such a world?

5. Write the opening of a story set during the last day of the universe. Pick whichever ending you want, but try to make your story fit your choice.

6. Time travel is always a popular idea. What would be some of the benefits of being able to travel in time? What would be some of the problems?

7. Think of a book or a movie that uses time travel. What sort of explanation is offered? Is it believable? Pick the book or movie that offers the most realistic time travel theory.

8. Alpha Centauri is 4.3 light-years from Earth. If you were orbiting Alpha Centauri and had a telescope so powerful you could see right into your house on Earth, what would you see?

9. We do not have any evidence that there is life elsewhere in our galaxy. However, we don't have any evidence that there isn't, either. Therefore, the existence of extraterrestrial civilizations is a matter of belief. What do you believe, and why?

ADDITIONAL ACTIVITIES

1. Additional History Dials (Individual) Consider the "Dial the Evolution of the Universe" activity on page 143. Follow the same general procedure to make a dial of the evolution of a star. (The activity can be adapted to anything that develops in stages that can be illustrated.)

2. Milky Way Poster (Individual) Make a poster model of the Milky Way galaxy. Indicate the position of the sun and any other reference points you can find through research.

3. Beam a Message (Individual or small group) Examine a copy of the message beamed into space from a radio telescope:

After setting up a code in the top row (counting from 1 to 10 in binary numbers), the message continues to include a description of hydrogen, a picture of DNA, a human figure, a view of the sun and its planets (with Earth highlighted), and a drawing of the radio telescope at Aricibo.

On a piece of graph paper, design your own message for extraterrestrials. You can only send two signals: either a solid square or an empty square. Make your message understandable and meaningful.

4. The Great Stellar Scavenger Hunt
(Individual or small group) Your job is to find objects that, by their designs or by their names, have to do with stars or galaxies. For example, a baseball team (Astros). As soon as you have everything on the list, you are done.

a car	a different flag item
a candy	a company
movie title	a television show
household product	a friend's name
fast food	a hardware store item
a song title	clothing
a flag item	(feel free to add more)

5. Time Story
(Creative, interdisciplinary) Create a short story where time travel plays a key part. Make sure your story has all the things a story should have: character, plot, setting, conflict, and conclusion. What sort of time travel theory did you use? Exchange and publish your stories.

Bogglers' Solutions
SORTING OUT THE NEBULAE
(page 136)

Matches:

Ring — Lyra — Planetary
Lagoon — Sagittarius — Emission
Horsehead — Orion — Dark
Dumbbell — Vulpecula — Planetary
Trifid — Sagittarius — Emission
Orion — Orion — Emission

(See chart on page 151.)

THE BIG BANG DOUBLE-BOGGLER
(page 138)

1. Albert Einstein
2. Vesto Slipher
3. Willem deSitter
4. Alexander Friedmann
5. Stephen Hawking
6. Arthur Eddington
7. Edwin Hubble
8. George Gamow
9. Henrietta Leavitt
10. Penzias & Wilson

Secret Message: Greetings from the Big-Bang Gang!"

	SAGITTARIUS	SAGITTARIUS	ORION	ORION	VULPECULA	LYRA	PLANETARY	PLANETARY	EMISSION	EMISSION	EMISSION	DARK
RING	X	X	X	X	X	O	O	X	X	X	X	X
LAGOON	O	X	X	X	X	X	X	X	O	X	X	X
HORSEHEAD	X	X	O	X	X	X	X	X	X	X	X	O
DUMBBELL	X	X	X	X	O	X	X	O	X	X	X	X
TRIFID	X	O	X	X	X	X	X	X	X	O	X	X
ORION	X	X	X	O	X	X	X	X	X	X	O	X
PLANETARY	X	X	X	X	X	O						
PLANETARY	X	X	X	X	O	X						
EMISSION	O	X	X	X	X	X						
EMISSION	X	O	X	X	X	X						
EMISSION	X	X	X	O	X	X						
DARK	X	X	O	X	X	X						

ADDITIONAL READING

Books

Asimov, Isaac. *Frontiers*. New York: Penguin Group Publishing, 1990.

Asimov, Isaac. *Frontiers II*. New York: Penguin Group Publishing, 1993.

Asimov, Isaac. *The Secret of the Universe*. New York: Windsor Publishing, 1992.

Barnes-Svarney, Patricia. *Traveler's Guide to the Solar System*. New York: Sterling Publishing, 1993.

Brenner, Barbara. *Planetarium*. New York: Bantam Books, 1993.

Comins, Neil. *What if the Moon Didn't Exist?* New York: HarperCollins Publishers, 1993.

Corrigan, Grace George. *A Journal for Christa*. Lincoln, NE: University of Nebraska Press, 1993.

Couper, Heather and Nigel Henbest. *Reader's Digest How the Universe Works*. Pleasantville, NY: Reader's Digest Association, Inc., 1994.

Dickinson, Terence. *The Big Bang to Planet X*. New York & Ontario: Camden House, 1993.

Dickinson, Terence. *Extraterrestrials: A Field Guide for Earthlings*. New York & Ontario: Camden House, 1994.

Harrington, Philip and Edward Pascuzzi. *Astronomy for All Ages*. Old Saybrook, CT: Globe Pequot Press, 1994.

Hawking, Stephen. *A Brief History of Time: A Reader's Companion*. New York: Bantam Books, 1992.

Miller, Ron and William Hartmann. *The Grand Tour: A Traveler's Guide to the Solar System*. New York: Workman, 1993.

Moore, Patrick. *Space Travel for Beginners, Astronomy for Beginners, The Universe for Beginners*. New York: Cambridge University Press, 1992.

Progue, William. *How Do You Go to the Bathroom In Space?* New York: Tor Books, 1991.

Ride, Sally and Susan Okie. *To Space & Back*. New York: William Morrow, 1989.

Tennant, Catherine. *The Box of Stars — A Practical Guide to the Night Sky and Its Myths & Legends*. Boston: Bulfinch Press, 1993.

Magazines

Discover, The World of Science. Walt Disney Magazine Publishing Group, Inc., 114 Fifth Avenue, New York, NY 10011. Phone: 800-829-9132.

Odyssey. Cobblestone Publishing, Inc., 7 School Street, Peterborough, NH 03458. Phone: 603-924-7209.

Sky & Telescope. Sky Publishing Corporation, 49 Bay State Road., Cambridge, MA 02138. Phone: 800-253-0245.

GUIDE TO INFORMATION AND RESOURCES

Computer Software

Broderbund Discover Space. Broderbund Software, Inc., P.O. Box 6125, Novato, CA 94948-6125. Phone: 415-382-4400.

Requirements: IBM PC-compatible computer, 560 KB of free memory, 7 MB of hard disk space, 3.5 floppy disk drive, DOS 3.1, and VGA display.

Description: Part computerized encyclopedia and part planetarium-type program. Great for beginners. The opening screen offers six different options: studying the sun, surveying the planets, exploring 18 deep-sky objects, examining comets and asteroids, learning about space exploration, and displaying a sky map.

Electronic Picturebooks: Space Telescope Science Institute Gems of Hubble; Space Telescope Science Endeavor Views the Earth; Space Telescope Science Institute Images of Mars; Space Telescope Science

Institute Magellan Highlights of Venus; Space Telescope Science Institute Scientific Results from GHRS. Space Telescope Science Institute, distributed by the Astronomical Society of the Pacific, 390 Ashton Avenue, San Francisco, CA 94112. Phone: 415-337-2624.

Requirements: Color Macintosh with System 7.X, 2.5 MB of memory, and a hard disk. It includes Hypercard Player 2.1.

Description: Multimedia presentations of color images, maps, informative text, and reference material intended for teachers, students, and armchair explorers.

Virtual Reality Distant Suns for Macintosh. Virtual Reality Laboratories, Inc., 2341 Ganador Court, San Luis Obispo, CA 93401. Phone: 805-545-8515.

Requirements: 2 MB of memory, System 6.0.2, and 3.6 MB free on hard disk. Macintosh II, math coprocessor, and 256-color system are recommended.

Description: It displays 9,824 stars and almost 2,000 galaxies, nebulae, and star clusters. Displays comets and asteroids, produce star charts, and show animations of eclipses.

Maris Redshift: Multimedia Astronomy. Maris Multimedia Ltd . Phone: 1-800-33-MAXIS

Requirements: CD-ROM for Windows and Macintosh.

Description: This CD-ROM is a collection of photos, text, diagrams, charts, and QuickTime clips of stars, constellations, galaxies, and the planets. The depth and quality of information make it a superb multimedia reference work for both students and teachers. Included are a series of NASA film clips, an on-line version of the *Penguin Dictionary of Astronomy*, and a virtual-reality capability that allows you to manipulate planetary objects.

Where in Space Is Carmen San Diego, school edition. Broderbund Software, Inc., P.O. Box 6125, Novato, CA 94948-6125. Phone: 415-382-4400.

Description: As with other *Where in the. . . * programs, something is stolen — in this case, the items relate to astronomy, such as the asteroid Vesta snatched from the asteroid belt. It is up to the students to deduce who of Carmen's gang is responsible. Students must use critical-thinking skills along with a basic knowledge of astronomy.

Asimov's the Ultimate Robot. Microsoft Corporation, Redmond, WA. 1993.

Requirements: IBM or Macintosh. A sound card and speakers are recommended.

Description: Information about robots and their role in our society. A workshop where you can assemble the robot of your dreams. Also includes stories by Asimov about robotics.

Cosmic Osmo and the World Beyond the Material. Cyan, Inc., Spokane, WA; distributed by Broderbund Software, Inc., 1994. Phone: 415-382-4400.

Requirements: Macintosh only (CD-ROM).

Description: Explore strange planets with bizarre sound effects and animation. For kids of all ages.

Computer Software Plus!

Connect with the 61-centimeter telescope at the Mt. Wilson Observatory in California from your home or school computer. Operated by the Mt. Wilson Institute, the Telescopes in Education program encourages the use of the telescope for science projects. Requirements include software and a modem plus $300 to $500 for a half-night or full-night of observation time. Software suggested is *Remote Astronomy* available from Software Bisque, 912 12th Street, Suite A, Golden, CO 80401. Phone: 303-278-4478. The software consists of three components and requires Microsoft Windows 3.1 and at least 4 MB RAM. Reservations for observing time can be made by contacting Gil Clark at Telescopes in Education, P.O. Box 24, Mt. Wilson, CA 91023. Phone: 818-395-7579. Fax: 818-793-4570.

Space-Related Organizations

Young Astronaut Council

(For elementary and middle school students) YAC is not a part of NASA; it is a private organization. YAC was established to provide direction for the creation of materials and activities for the Young Astronaut Program. Although its aim is to stimulate interest in science, math, and technology, YAC considers its materials to be a motivator in all subject areas. Chapters may be formed in schools and communities.

The Young Astronaut Council, P.O. Box 65432, Washington, DC 20036. Phone: 202-682-1984. Fax: 202-775-1773

Challenger Center for Space Science Education

(For students and teachers grades 5–8) Workshops, learning centers, and simulated missions are among the programs.

Challenger Center for Space Education, 1101 King Street, Suite 190, Alexandria, VA 22314. Phone: 703-683-9740

National Air and Space Museum's Educational Resource Center

(For teachers K–12) Provides lesson plans, videotapes, computer software and periodicals.

Educational Resource Center, Office of Education P-700, National Air & Space Museum, Smithsonian Institution, Washington, DC 20560. Phone: 202-786-2106

U.S. Space Camp

U.S. Space Camp has facilities in Alabama and Florida and has plans to open more.

U.S. Space Camp, One Tranquility Base, Huntsville, AL 35807. Phone: 800-637-7223

Additional Education Programs and Teacher Resource Centers

Each state has a regional educational resource center. For information contact the following national centers:

Residents of Alaska, Arizona, California, Hawaii, Idaho, Montana, Nevada, Oregon, Utah, Washington, Wyoming, contact: Teacher Resource Center, NASA Ames Research Center, Mail Stop TO25, Moffett Field, CA 94035. Phone: 415-604-3574

Residents of Connecticut, Delaware, District of Columbia, Maine, Maryland, Massachusetts, New Hampshire, New Jersey, New York, Pennsylvania, Rhode Island, Vermont, contact:Teacher Resource Laboratory, NASA Goddard Space Flight Center Mail Code 130.3, Greenbelt, MD 20771. Phone: 301-286-8570

Residents of Colorado, Kansas, Nebraska, New Mexico, North Dakota, Oklahoma, South Dakota, Texas, contact: Teacher Resource Room, NASA Johnson Space Center, Mail Code AP-4, Houston, TX 77058. Phone: 713-483-8696

Residents of Florida, Georgia, Puerto Rico, Virgin Islands, contact: Educators Resources Laboratory, NASA John F. Kennedy Space Center, Mail Code ERL, Kennedy Space Center, FL 32899. Phone: 407-867-4090

Residents of Kentucky, North Carolina, South Carolina, Virginia, West Virginia, contact: NASA Teacher Resource Center, Virginia Air and Space Center, 600 Settler's Landing Road, Hampton, VA 23669-4033. Phone: 804-727-0900, x757

Residents of Illinois, Indiana, Michigan, Minnesota, Ohio, Wisconsin, contact: NASA Lewis Research Center, Mailstop 8-1, 21000 Brookpark Road, Cleveland, OH 44135. Phone: 216-433-2017

Residents of Alabama, Arkansas, Iowa, Louisiana, Missouri, Tennessee, contact: U.S. Space & Rocket Center, NASA Teacher Resource Center for MSFC, Huntsville, AL 35807. Phone: 205-544-5812

Residents of Mississippi, contact: NASA Stennis Space Center, Teacher Resource Center, Building 1200, Stennis Space Center, MS 39529. Phone: 601-688-3338

Glossary

ASTROPHYSICS — the study of the energy emitted from stars

CHROMATIC ABERRATION — the separation of the colors of white light as it is focused by a lens

CHROMOSHPERE — the layer of the sun's atmosphere surrounding the photosphere

CONCAVE — a shape (or lens) that is narrower at the center than at the edge

CONSTELLATION — a group of stars that seems to form some pattern in the night sky

CONVEX — a shape (or lens) that is wider at the center than at the edge

CORONA — the outermost layer of the sun's atmosphere, visible during a total solar eclipse

COSMOLOGY — the study of the universe, its origin, and its evolution

DARK MATTER — matter that is not visible by visual, infrared, or other means of detection

DOGMA — beliefs that one is expected to accept without question

EPICYCLE — a small circle that moves along a larger circle (i.e., the orbit of the moon as it also orbits with the Earth around the sun)

ESCAPE VELOCITY — the speed necessary to break free from an object's gravitational pull

EVENT HORIZON — the outer "edge" of a black hole, where the escape velocity is exactly equal to the speed of light

FUSION — a nuclear reaction that causes a joining of nuclei and a release of electrons

GEOCENTRIC THEORY — a theory placing the Earth at the center of the solar system

GEOTHERMAL — a form of energy that is based on the internal heat of the Earth

GRAVITY — the force of attraction that one object exerts upon another

GREENHOUSE EFFECT — a heating of the atmosphere as a result of trapped radiation

HEEL STONE — a key marker stone at primitive observational sites

HELIOCENTRIC THEORY — a theory that places the sun at the center of the solar system

HOROSCOPE — a prescription for daily living based on birth date and the positions of stars and planets

HYPOTHESIS — a theory tested under controlled conditions

KELVIN KRUSH — the excruciating headache you get when you suck too long on a snow cone

ISOTOPE — a closely related form of an element, varying in atomic weight or nucleus

LOCAL GROUP — the galaxies and clusters that move through space together with the Milky Way and Andromeda

LUNAR ECLIPSE — a phenomenon occurring when the moon enters the shadow cast by the Earth

MAGMA — molten material from within the Earth's crust

MAGNETOSPHERE — the extent of influence of the magnetic field of a planet or star

MAGNITUDE — an indication of the brightness of a star or other object

MEDICINE WHEEL — a Native American design on the earth serving as an almanac and calendar

MICROGRAVITY — the condition of apparent weightlessness when an object is at some distance from a center of gravity

NEBULA — an interstellar area of relatively dense dust and gases

NEUTRON — a stable elementary particle found in the nucleus of the atom and carrying no charge

NOVA — a stellar "explosion," casting stellar matter into space

PANGAEA — the name of the single continent that later broke apart on separate tectonic plates

PHOTOSPHERE — the visible surface of the sun (or the boundary between the interior and exterior of any star)

PRECESSION — a slow drifting or cyclical change, particularly of the axis of the Earth

PULSAR — a neutron star that rotates at a rapid and regular rate, emitting radiation as it rotates

QUASAR — a "quasi-stellar" source of tremendous energy, perhaps the exploding nucleus of a galaxy

RED SHIFT — an aspect of the Doppler effect, where key markers on spectra of receding stars shift to the red side

REHYDRATION — to add moisture a second time

SCIENTIFIC METHOD — a controlled experimental evaluation of a hypothesis

SOLAR ECLIPSE — a phenomenon that occurs when the moon passes in front of the sun, casting a shadow on a narrow band of the Earth

SOLAR FLARE — an eruption in the outer part of the sun's atmosphere

SOLAR PROMINENCE — large masses of glowing gas from the sun's corona

SOLAR WIND — ionized gas emitted by the sun

SOLSTICE — The point at which the sun has reached the northernmost or southernmost point on the ecliptic

SPECTROSCOPY — a scientific study of the spectra of various energy sources

SUBDUCTION — the forcing of one tectonic plate into the magma as it comes in contact with another tectonic plate

SUNSPOT — a relatively cool and dark spot on the sun's photosphere

SUPERNOVA — a tremendous explosion of the outer layers of a star, caused perhaps by the gravitational collapse of the star's core

SUPRELLA HARRISON — nobody in particular

TERRESTRIAL — having to do with the Earth

WORMHOLE — a theoretical device that would bend space so drastically as to create "shortcuts" between vast distances

Charts and Tables

Do You Measure Up?

. . . for starters . . .

1 inch	=	2.54 centimeters	-or-	1 centimeter	=	0.394 inches
1 foot	=	0.305 meters	-or-	1 meter	=	3.3 feet
1 yard	=	0.915 meters	-or-	1 meter	=	1.094 yards
1 mile	=	1.609 kilometers	-or-	1 kilometer	=	0.621 miles
1 pound	=	448 grams	-or-	1 kilogram	=	2.205 pounds

(and by the way) 1 ton = 2240 pounds

. . . and for the number-crunchers .

1 A (angstrom)	=	.0000000001 (one ten-billionth) meters
1 year	=	365.242199 days
1 AU (astronomical unit)	=	149,597,870 kilometers
SOL (speed of light)	=	299,792,458 km per second
1 LY (light year)	=	9.46 trillion km
1 parsec	=	3.262 LY -or- 2062265 AU

6.71 meters = the full length of the Giant Earthworm from South Africa (just in case you were wondering)

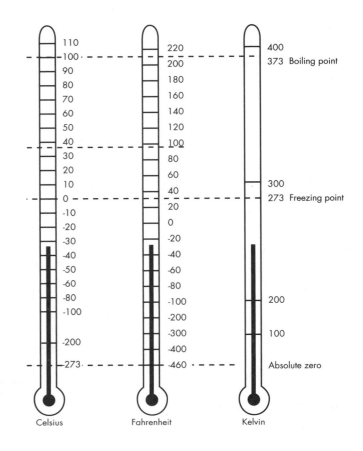

Celsius   Fahrenheit   Kelvin

373 Boiling point
273 Freezing point
Absolute zero

The Electromagnetic Spectrum

High Frequency   Low Frequency

Short Wavelength   Long Wavelength

Gamma rays	X-Rays	Ultraviolet rays	VISIBLE LIGHT	Infrared rays	Microwaves	Radio Waves

cancer treatment, medicine, suntan, photosynthesis, heat, cooking, radar, satellite comm., FM radio, television, short wave radio, radio astronomy, AM radio

Future Eclipses

Future Total Solar Eclipses

1995 (October 24) Indonesia and Southeast Asia

1997 (March 9) Russia

1998 (February 26) South America, Caribbean, Pacific

1999 (August 11) Europe, Middle East, India

2001 (June 21) Africa, South Atlantic

Future North American Partial Eclipses

1998 (February 26) all but west coast and northwest

2000 (December 25) northeast

2001 (December 14) west

2002 (June 10) west

Future North American Total Eclipses

1996 (April 3) eastern half

1996 (September 26) all of North America

2000 (January 20) all of N.A.

2003 (May 15) all of N.A.

2003 (November 8) all but west coast

2004 (October 27) all of N.A.

PLANET	AVG. KM. FROM SUN	AU	MASS (Earth=1)	DIAMETER (km)	GRAV. (Earth=1)	ESC. VEL. (km/sec)	KNOWN MOONS	ORBIT PERIOD	ROTATION PER.
Mercury	57910000	0.387	0.554	4850	0.377	4.3	0	88 days	59 days
Venus	108200000	0.7233	0.815	12140	0.905	10.4	0	225 days	244 days
Earth	149579870	1	1	12756	1	11.2	1	365.26 days	23h 56m 4s
Mars	227940000	1.5237	0.107	6790	0.379	5.3	2	1.88 years	24h 37m 23s
Jupiter	778320000	5.2028	317.89	142800	2.54	60.2	16	11.86 years	avg. 9h 53m
Saturn	1427000000	9.5388	95.26	120660	1.07	32.3	20	29.45 years	avg. 10h 25m
Uranus	2869600000	19.1819	14.6	51400	0.86	22.5	15	84.07 years	12h to 24 h
Neptune	4496600000	30.0578	17.2	49530	1.14	23.9	8	164.81 years	16h 3m
Pluto	5913708900	39.53	0.0026	2253	0.03	?	1	248.5 years	6days 9h 17m

Bibliography

Anderson, Charlene M. "Galileo Encounters Earth and Venus." *The Planetary Report*, March/April 1991, pages 12–15.

Anderson, John. "Planet X: Fact or Fiction." *The Planetary Report*, July/August 1988, pages 6–9.

Atkinson, Stuart. *Journey into Space*. New York: Viking Penguin, Inc., 1988.

Beatty, J. Kelly and Andrew Chaikin. *The New Solar System*. Cambridge, MA: Sky Publishing, 1990.

Becklake, Sue. *Space, Stars, Planets and Spacecraft*. New York: Dorling Kindersley, 1991.

Burnham, Robert. "Something from Nothing." *Astronomy*, December 1994, page 6.

Cuzzi, Jeffrey N. "Saturn: Jewel of the Solar System." *The Planetary Report*, July/August 1989, pages 12–15.

Diagram Group. *Comparisons*. New York: St. Martin's Press, 1980.

Dickinson, Terence. *From the Big Bang to Planet X*. New York: Camden House, 1993.

Dickinson, Terence. "Black Holes Ain't So Black." *Odyssey*, January 1995, page 15.

Dickinson, Terence. "Black Holes, a.k.a. Cheshire Cats." *Odyssey*, January 1995, pages 10–15.

Dowling, Claudia Glenn. "This Precious Planet." *Life*, April 1992, pages 30–36.

El Khazen, Neda. "Interview: Trinh Xuan Than," *UNESCO Courier*, May 1994, pages 4–7.

Ferris, Timothy. *Coming of Age in the Milky Way*. New York: Doubleday, 1988.

Flamsteed, Sam, "The Great Comet Crack–Up." *Discover*, January 1995, pages 28–32.

Gallant, Roy. *National Geographic Picture Atlas of Our Universe*. Washington, DC: National Geographic Society, 1986.

Gatland, Kenneth. *The Illustrated Encyclopedia of Science Technology*. New York: Orion Books, 1981.

Gustafson, John. *Stars, Clusters and Galaxies*. New York: Julian Messner, 1992.

Halpern, Paul. *Cosmic Wormholes*. New York: Penguin Group, 1993.

Hartmann, William and Ron Miller. *The History of the Earth*. New York: Workman, 1991.

Hatchett, Clint. *Discover Planetwatch*. New York: Running Heads, Inc., 1993.

Hausman, Gerald. *Tunkashila*. New York: St Martin's Press, 1993.

Hawking, Stephen, W. *A Brief History of Time*. New York: Bantam Books, 1988.

Henbest, Nigel. *The Planets*. London: Penguin Books, 1992.

Horowitz, Paul. "Project META — What Have We Found." *The Planetary Report*, September/October 1993, pages 4–9.

Jastrow, Robert. *Red Giants and White Dwarfs*. New York: W.W. Norton & Company, 1990.

Kasting, James F. "Earth, the Living Planet: How Life Regulates the Atmosphere." *The Planetary Report*, January/February 1990, pages 8, 9, 24.

Lauber, Patricia. *Seeing Earth from Space*. New York: Orchard Books, 1990.

Levy, David H. "The Man Who Found Pluto." *Odyssey*, May 1992, pages 14–17.

Matthews, Robert. "Cosmologists Meet to Face Their Fears." *Science*. November 5, 1993, pages 846–847.

McAleer, Neil. *The Cosmic Mind–Boggling Book*. New York: Warner Books, 1982.

McAleer, Neil. *The Mind–Boggling Universe*. New York: Doubleday, 1987.

McKay, David, W. and Bruce G. Smith. *Space Science Projects for Young Scientists*. New York: Franklin Watts, 1986.

Menzel, Donald, H. *The Random House Illustrated Science Library — Astronomy*. New York: Random House, 1975.

Miner, Ellis D. "Voyager 2 Approaches Neptune." *The Planetary Report*, July/August 1989, pages 19–21.

Moeschl, Richard. *Exploring the Sky*. Chicago: Chicago Review Press, 1993.

Moore, Dianne, F. *The Harper Collins Dictionary of Astronomy and Space Science*. New York: HarperCollins, 1992.

Morrison, David. "Jupiter: First Stop on Voyager's Grand Tour." *The Planetary Report*, July/August 1989, pages 8–11.

Nicolson, Iain. *The Illustrated World of Space*. New York: Simon & Schuster Books for Young Readers, 1991.

Ordway, Frederick L. III and Randy Liebermann, editors. *Blueprint for Space*. Washington, DC: Smithsonian Institution Press, 1992.

Pasachoff, Jay, M. *Peterson First Guides — Astronomy*. Boston: Houghton Mifflin Company, 1988.

Pivirotto, Donna. "Rovers! Using Mobile Robots as Planetary Explorers." *The Planetary Report*, July/August 1991, pages 8–13.

Reid, Struan. *The Usborne Book of Space Facts*. Tulsa, OK: Educational Developmental Corporation, 1987.

Sagan, Carl. *Cosmos*. New York: Random House, 1980.

Sagan, Carl. *The Pale Blue Dot*. New York: Random House, 1994.

Sagan, Carl. "Exploring Other Worlds and Protecting This One: The Connection." *The Planetary Report*, January/February 1990, pages 4–7.

Sagan, Carl and Ann Druyan. "Snowflakes Fallen on the

Hearth: The Evolution of the Earth." *The Planetary Society*, January/February 1993, pages 4–9.

Simon, Seymour. *Comets, Meteors, and Asteroids*. New York: Morrow Junior Books, 1994.

Simpson, Richard A. and Ellis D. Miner. "Uranus: Beneath That Bland Exterior." *The Planetary Report*, July/August 1989, pages 16–18.

Soffen, Gerald. "Biosphere 2, A Living Experiment." *Odyssey*, March 1993, pages 34–37.

Soffen, Gerald. "Spinoffs." *NASA Publications*, 1988.

Trefil, James. "Stephen Hawking and Quantum Gravity." *Odyssey*, January 1995, pages 19–24.

VanCleave, Janice. *Astronomy for Every Kid*. New York: John Wiley and Sons, Inc., 1991.

Villard, Ray. "Expanding Our View of the Universe." *Odyssey*, June 1993, pages 18–24.

Wetterau, Bruce. *The New York Public Library Book of Chronologies*. New York: Prentice Hall Press, 1990.

Wilford, John Noble. "Science Times: New Puzzle Arises on Universe's Age." *New York Times*, October 4, 1994, page 11, col. 1.

Wood, Robert. *Science for Kids*. Blue Ridge Summit, Pennsylvania: Tab Books, 1991.

Author Biography

Steven R. Wills brings his interest in astronomy to *ODYSSEY* magazine, where he has written the "Mind Bogglers" feature for more than five years. Although *Mind-Boggling Astronony* is his first full-length book, his writing credits include articles for a variety of magazines and a weekly newspaper column on education.

He also has taught high school English for 24 years, where he manages to pull in references to astronomy whenever he can get away with it. (Let's see...didn't Mark Twain "come in" and "go out" with Halley's comet?)

Index

Aristarchus, heliocentric theory, 8

Aristotle, geocentric theory, 17

Asteroid Belt, 50-52
    Apollo asteroids, 51
    Arizona Crater, 52
    extinction of dinosaurs, 52, 53
    Guiseppi Piazzi, 51
    Johanne Bode, 51
    Tunguska, Siberia, 52

Babylonians
    horoscopes, 11
    solar eclipse, 8, 12
    zodiac constellations, 11, 12

Big Bang, 137, 139-142
    COBE, 137
    cosmic timeline, 139-142

Black holes, 128, 129

Brahe, Tycho, 19-21

Calendars, 9-15
    Aztec, 14, 15
    Egyptian, 9
    Great Pyramid of Giza, 13
    Julian, 9
    North American Indian medicine
        wheels, 14
    starting points, 15

Cameras and Imaging, 96-98
    spectroscopy and stellar classification,
        98

Chinese
    first recorded comet, 8, 12
    lunar eclipse, 12
    supernova and sunspots, 12

Copernicus, 18, 19
    De Revolutionibus, 19
    earth's rotation, 19
    heliocentrism, 19

Cosmology, 123

Dark matter, 142

Doppler Effect, 136

Earth, 74-116
    atmosphere, 46, 47; 80, 81
    geologic history, 83
    Greenhouse Effect, 85, 86
    interior, 75-76
    ozone, 86-88
    relationship with moon, 38-39
    surface, 77-80

Einstein, Albert, 123

Galaxies, 131-133, 135
    Edwin Hubble and the red shift,
        135-136
    local group, 132

Galileo, 23-26
    acceleration, 24
    Dialogue, 26
    Inquisition, 25-26
    pendulum clock, 24
    refracting telescope, 24, 25

Hawking, Stephen, 145, 146

Hubble Space Telescope, *see telescopes*

Jupiter, 53-56
    composition, 54
    energy and mass, 54
    Great Red Spot, 54
    magnetosphere, 53-54
    moons (Callisto, Ganymede, Europa,
        Io), 55, 56
    Shoemaker-Levy comet collision, 54,
        55

Kepler, Johannes, 19-21
    effect of tides, 20
    elliptical orbits, 20
    Laws of Planetary Motion, 20, 21

Mars, 42-50
    atmosphere, 46
    surface, 44, 45
    terraforming, 48-50

Mercury, 42-47
    atmosphere, 45, 46
    surface, 44

Moon, 38-40; 43, 44
    composition, 38-40
    Earth-moon relationship, 39
    impact cratering, 43, 44
    origin, 39
    tides, effect of, 40

Mythology, 9-11
    Egyptian, 9
    Greek and Roman, 10
    Native American, 10

Neptune, 60-62
    composition, 61
    moons, (Triton, Phoebe), 61

Newton, Isaac, 23-26
    Laws of Celestial Dynamics, 26
    Principia, 26
    reflecting telescope, 25
    theory of gravitation, 25

Planets, *see individual planet names*

Phoenicians, celestial navigation, 12

Pluto, 62, 63
    Charon relationship, 63
    Kuiper Belt, 63
    Oort Cloud, 63, 64
    Tombaugh, Clyde, 62, 97

Ptolemy, 17, 18
    Almagest, 17
    geocentric theory, 17
    magnitudes, 18
    precession of the equinox, 18

Quasars, 146

Radio astronomy, *see telescopes*

Rockets and robots, 100, 101, 103, 104
    Robert Goddard, 101
    Mars rover, 103, 104
    satellites, 104
    Konstantin Tsiolkovsky, 100, 101
    von Braun, Wernher, 101
    Voyager missions, 103

Satellites, *see rockets and robots*

Saturn, 57-59
    composition, 58
    energy, 58
    moons (Mimas, Dione, Hyperion,
        Titan), 59
    rings, 58, 59

Search for extraterrestrial life, 147

Space shuttle, 107-111

Space station, 111-113

Stars, 124-129
    birth, 124-126
    death, 127-129

Stonehenge: 12-14

Sun, 33-35
    composition, 33, 34
    energy from, 35
    Kelvin Scale, 34

**Telescopes and Observatories, 89-91;
    93-95, 98**
    Hubble Space Telescope, 95
    radio telescopes, 98

Time, 144-146

Uranus, 60, 61
    axis of rotation, 61
    composition, 60, 61
    moons (Miranda), 61
    rings, 61

Venus, 42-47
    atmosphere, 46
    Greenhouse Effect, 46
    surface, 45

***Voyager 1* and *2*,** *see rockets and robots*

Wormholes, 146